"PENETRATING"
—Science Digest

"The future of the universe: its possible collapse, quasars as 'white holes,' space-time 'wormholes,' and (the) ultimate recycling of the universe . . . The prolific **Dr. Asimov,** that one-man publisher's list, has done it again."
—Smithsonian Magazine

"Stimulating and cogent . . . **Asimov proves his reputation as the dean of popular science writers.** The most abstract and difficult concepts are made readily comprehensible in prose as lucid as it is informative."
—The Los Angeles Times

"Those with patience to follow Asimov's lead will discover that they can grasp the significance of **the most exciting discoveries in astronomy in recent decades.**"
—St. Louis Post-Dispatch

Books by Isaac Asimov

Asimov on Numbers
The Beginning and the End
The Collapsing Universe

Published by POCKET BOOKS

ISAAC ASIMOV

THE COLLAPSING UNIVERSE

PUBLISHED BY POCKET BOOKS NEW YORK

POCKET BOOKS, a division of Simon & Schuster, Inc.
1230 Avenue of the Americas, New York, N.Y. 10020

Published by arrangement with Walker & Company
Library of Congress Catalog Card Number: 76-53639

ISBN: 0-671-49886-X

First Pocket Books printing April, 1978

10   9   8   7

POCKET and colophon are registered trademarks
of Simon & Schuster, Inc.

Printed in the U.S.A.

CONTENTS

1 PARTICLES AND FORCES

SINCE 1960 the universe has taken on a wholly new face. It has become more exciting, more mysterious, more violent, and more extreme as our knowledge concerning it has suddenly expanded. And the most exciting, most mysterious, most violent, and most extreme phenomenon of all has the simplest, plainest, calmest, and mildest name—nothing more than a "black hole."

A hole is nothing, and if it is black, we can't even see it. Ought we to get excited over an invisible nothing?

Yes—if that black hole represents the most extreme state of matter possible, if it represents the possible end of the universe, if it represents the possible beginning of the universe, if it represents new physical laws and new methods for circumventing what had previously been considered absolute limitations.

In order to understand the black hole, however, we must begin at the beginning and work our way toward it step by step.

THE FOUR FORCES

There are four different ways in which the various particles that make up the universe can interact with one another. Each of these is a particular variety of *interaction*, or to use a more old-fashioned but more common term, a *force*. Scientists have been unable to detect a fifth force, or yet, to find any reason why a fifth should be required.

The four forces are listed in order of decreasing strength in Table 1.

TABLE 1—Relative Strength of the
Four Forces

Force	Relative Strength *
Nuclear	10^3
Electromagnetic	1
Weak	10^{-11}
Gravitational	10^{-39}

Every particle in the universe is the source of one or more of these forces. Each particle serves as the center of a volume of space in which that force exists with an intensity that decreases as the distance from the source increases. The volume of space in which that force can make itself felt is the *force field*.

Any particle that can serve as the source of a particular field will respond to such a field set up by another particle. The response is generally one

---

* The relative strengths are given in exponential numbers, where 10^3 stands for 1,000 and 10^{-11} for 1/100,000,000,000. Some details concerning exponential numbers are given in Appendix 1 if you're not familiar with their use.

of movement: the particles moving toward each other (an *attraction*) or away from each other (a *repulsion*) unless physically constrained from doing so.

Thus, any object capable of producing a gravitational field will, if placed in Earth's gravitational field, move toward Earth's center—that is, it will fall. The Earth will also move toward the object's center, but since it will likely be much larger than the falling object, it will rise correspondingly more slowly—usually, in fact, immeasurably slowly.

Of the four forces two—the *nuclear force* and the *weak force*—make themselves felt only at incredibly tiny distances of 10^{-13} centimeters or less.* This is just about the width of the tiny nucleus that exists at the very center of the atom. It is only within the nucleus, in the immediate neighborhood of isolated particles, that these forces exist. For this reason the term *nuclear force* is sometimes given to both, and they are differentiated by their relative strength into the *strong nuclear force* and the *weak nuclear force*.

In this book, however, there will be little occasion to mention the weak force, so we will simply refer to the stronger force as the nuclear force and be done with it.

A given particle is not likely to produce and to respond to each of the forces. Only certain particles, for instance, produce and respond to the nuclear force. Those that do are called *hadrons*, from a Greek word meaning "strong," since the nuclear force is the strongest of the four. The

---

* In this book I am using the metric system, which is just about universally used outside the United States and is used by scientists within the United States, too. Some details concerning the metric system are given in Appendix 2.

hadrons that are most common and most important to the structure of the universe are the two
nucleons—the *protons* and the *neutron*.

The proton was discovered in 1914 by the Brit-
ish physicist Ernest Rutherford (1871-1937), and
its name comes from the Greek word for "first"
because at the time of its discovery it was the
smallest object known to have a positive electric
charge.

The neutron was discovered in 1932 by the
English physicist James Chadwick (1891-1974).
It carries no electric charge, either positive or
negative. In other words, it is electrically neutral;
hence its name.

As early as 1911 Rutherford had shown that
an atom contains almost all its mass in a very
small region at the center, the *nucleus*. Once protons were discovered, it was realized that they are
relatively massive particles and must be located
in the nucleus. The number of protons varies
from one kind of atom to another. The hydrogen
atom has a single proton in its nucleus, the helium
atom has 2, the lithium atom has 3, and so on—
up to the uranium atom which has 92. Still more
massive atoms have been prepared in the laboratory.

But what holds all the protons together in the
nucleus, where they are all squeezed into such
close proximity?

Prior to 1935 only two forces were known—the
electromagnetic and the gravitational. The *gravitational force* is too weak to hold the protons together. The *electromagnetic force* is strong
enough, but it can manifest itself as either an
attraction or a repulsion. Between two particles
of opposite electric charge (plus and minus) there

is an attraction. Between two particles of the same electric charge (plus and plus, or minus and minus) there is a repulsion. Protons are all positively charged and must therefore repel each other; and the repulsion must be more intense the closer the protons are to each other. In an atomic nucleus, with protons squeezed together until they are virtually in contact, the electromagnetic repulsion must be enormously strong—yet the protons hold together.

In addition to protons neutrons are also present in the nucleus, but this doesn't seem to help the situation. Since neutrons lack an electric charge, they neither produce nor respond to an electromagnetic force. They should therefore neither attract nor repel protons. They should neither help hold the protons together nor accelerate the breakup.

It was not until 1935 that the Japanese physicist Hideki Yukawa (1907-) put forth a successful theory of the nuclear force. He showed that it was possible for protons and neutrons when very close to each other to produce an attracting force a thousand times greater than the electromagnetic repelling force. What the nuclear force holds together the electromagnetic force cannot blow apart.

The nuclear force works best and keeps the nucleus stable only when protons and neutrons are present in certain proportions. For atoms whose nuclei contain 40 particles or less the best proportion seems to be equal numbers of protons and neutrons. For more complicated nuclei there has to be a preponderance of neutrons, that preponderance growing greater as the nucleus grows more complex. A bismuth nucleus, for instance, contains 83 protons but 126 neutrons.

When an atomic nucleus is forced to have proportions outside the region for stability, it does not remain intact. Small *beta particles* (beta is the second letter of the Greek alphabet) are given off under the influence of the weak force until the proportion is adjusted to stability. Other ways of nuclear breakup are also possible, but all these ways are lumped under the heading of *radioactivity*.

Strong as the nuclear force is, it has its limits. The intensity of the nuclear force falls off *very* rapidly with distance, and it can't make itself felt outside the nucleus. In fact its attractive influence fades considerably when it must extend itself from one end to the other of one of the larger nuclei.

The electromagnetic force also fades off, but much more slowly. The size of the nucleus is limited, since eventually the electromagnetic repulsion from end to end will equal the rapidly fading nuclear attraction from end to end. That is why atomic nuclei are so infratiny. The nuclear force simply won't produce anything larger (except under the most unusual conditions, which we will come to later in the book.)

Now let's concentrate on the electromagnetic interaction, which, as I have said, is produced only by those particles that carry an electric charge, and which is responded to only by those charged particles. The charge comes in two varieties, positive and negative. The force between positive and negative is an attraction, while that between positive and positive or between negative and negative is a repulsion.

The proton, with its positive electric charge, is a source of and responds to both the nuclear force and the electromagnetic force. The neutron, which is electrically uncharged, is a source of and responds to only the nuclear force.

Then, too, there are particles called *leptons*, from a Greek word meaning "weak," which are a source of and respond to the weak force but never to the nuclear force. Some leptons, however, are electrically charged, and they are a source of and respond to the electromagnetic force as well as to the weak force.

The most important of the leptons, as far as ordinary matter is concerned, is the *electron*, which carries a negative electric charge. (The beta particles produced by unstable nuclei by way of the weak force proved to be speeding electrons.) The electron was discovered in 1897 by the English physicist Joseph John Thomson (1856-1940), and it received its name because it was the smallest unit of electric charge then known (or for that matter, known today).

The information we now have can be summarized as in Table 2.

TABLE 2—Particles and Forces

	Proton	Neutron	Electron
Nuclear force	Yes	Yes	No
Electromagnetic force	Yes	No	Yes

NOTE: There also exist particles like the electron but with a positive electric charge. These are *antielectrons*, or *positrons*. A proton with a negative electric charge is an *antiproton*. A neutron with certain other properties reversed is an *antineutron*. As a group these opposites are *antiparticles*. Just as ordinary particles go to make up the matter all about us, antiparticles could make up *antimatter*. Such antimatter may exist somewhere in the universe, but we have never been able to detect it. Scientists can make tiny quantities of it in the laboratory, however.

ATOMS

Since electrons are not subject to the nuclear force, they cannot form part of the nucleus. Nevertheless, an electron is attracted to a proton, thanks to the electromagnetic force, and tends to remain near one. Thus, if a nucleus is made up of a single proton, there is likely to be a single electron held in its vicinity by the electromagnetic force. If there are two protons in the nucleus, there are likely to be two electrons held in its vicinity, and so on.

The nucleus and the nearby electrons make up the *atom*. (*Atom* is from a Greek word meaning "unbreakable" because at the time atoms were first dealt with, it was thought they could not be broken up into smaller units.)

As it happens, the charge on the electron is precisely equal (though opposite in nature) to the charge on the proton. Therefore, when there are x protons in the nucleus, the existence of x electrons in the regions just outside the nucleus will mean that the two kinds of charge will exactly neutralize each other. The atom as a whole is electrically neutral.

Although the electron and proton are equal in size of electric charge, they do not have the same mass.* The proton is 1836.11 times as massive as

---

* When we say that an object possesses mass, we mean that it takes force to make it move if it is standing still, or to change its speed or direction of movement if it is already moving. The more mass it has, the more force it takes. Under ordinary circumstances here on the surface of the Earth massive objects impress our senses as being "heavy." The more massive they are, the heavier. Still, mass and weight are not identical, and while the meaning is clear if we say that the proton is much heavier than the electron, it is safer to say "more massive."

the electron. Imagine an atom, then, with 20 protons and 20 neutrons in the nucleus and 20 electrons in the outer regions of the atoms. The electric charge is balanced, but more than 99.97 percent of the mass of the atom is in the nucleus.

Yet though the nucleus contains almost all the mass of an atom, it makes up only a tiny fraction of the volume of an atom. (This is an important point as far as the subject matter of this book is concerned—as we shall see.) The diameter of a nucleus is about 10^{-13} centimeters, while that of an atom is about 10^{-8} centimeters.

This means that an atom is 100,000 times as wide as a nucleus is. It would take 100,000 nuclei, placed side by side, to stretch across the atom of which it is part. If you imagine an atom to be a hollow sphere and start filling it with nuclei, it turns out that it would take 10^{15} (a million billion) nuclei to fill the atom.

Now let us consider two atoms. Each one has an overall electric charge of zero. We might suppose, then, that they would not affect each other, that they would be, so to speak, unaware of each other's existence, as far as the electromagnetic force was concerned.

Ideally, that would be so. If, in various atoms, the charge of the electron were spread with perfect evenness in a sphere about the nucleus, and if the positive charge of the nucleus were evenly mingled with the negative charge of the electrons, then the electromagnetic force would play no role between atoms.

That, however, is not the way it is. The negative charge of the electrons is present in the outer regions of the atom, and the positive charge of the nucleus is hidden within. When two atoms approach each other, it is the negatively charged outer region of one that is approaching the negatively charged outer region of the other. The two negatively charged regions repel each other (like charges repel), and that means that two atoms can only come so close to each other before they veer away or bounce off.

A sample of helium gas, for instance, is made up of separate helium atoms forever moving around and bouncing off each other. At ordinary temperature the helium atoms move quite rapidly and bounce off one another with considerable force. As the temperature lowers, however, the atoms move more and more slowly and rebound from each other more and more weakly. The atoms of the helium gas fall closer together, and the helium contracts and grows smaller in volume.

In reverse, as the temperature moves higher, the atoms move more quickly, rebound with greater force, and the helium expands.

There would seem to be no limit to how fast atoms could move (within reason), but there is an easy limit to how slowly they would move. If the temperature drops far enough, a point is reached where they move so slowly that no more energy can be withdrawn from them. At that level of frigidity we reach a temperature of *absolute zero*, which is −273.18°C.*

Although the helium atom has a charge distribution that is quite close to being perfectly symmetrical, it is only *quite* close and is not com-

* In this book temperature is expressed by means of the Celsius (centigrade) scale, which is used just about everywhere in the world but in the United States and by scientists even in the United States. For details on how that scale compares with the Fahrenheit scale, which is more familiar to Americans, see Appendix 3.

pletely perfect. The electric charge is not exactly evenly spread, and as a result, parts of the atom's surface are a little less negatively charged than others. As a result the atom's inner positive charge peeps through the less negative areas of the outside, so to speak, and two neighboring atoms will attract each other very weakly. This weak attraction is called the *van der Waals forces* because it was first worked out by the Dutch physicist Johannes Diderik van der Waals (1837-1923). As the temperature drops and helium atoms move more and more slowly, the force of rebound is eventually not great enough to overcome the tiny van der Waals forces. The atoms stick together, and helium gas becomes helium liquid.

The van der Waals forces are so weak in the highly symmetrical helium atom that the temperature must drop to as low as 4.3 degrees above absolute zero for helium liquid to form. All other gases have less symmetrical distribution of charge on their atoms; they therefore experience larger van der Waals forces and liquefy at higher temperatures.

Atoms can sometimes attract each other in stronger fashion. The electrons in the outer regions of the atoms are arranged in shells, and the structure is most stable if all the shells are filled. Except in the case of helium and a few similar elements atoms generally have their outermost shell not quite filled, or have a few surplus electrons left over when that shell is filled.

There is a tendency, then, for two atoms, on colliding, to transfer one or two electrons from the one that has extra to the one that has deficiency, which leaves both with filled outermost shells. But then the one that gains electrons has gained a negative charge, and the one that loses electrons can no longer balance the charge of its nucleus completely and has gained a positive charge. The two atoms then have a tendency to cling together.

Or else two atoms, on colliding, share electrons, which then help fill the outermost shell in both atoms. Both atoms then have filled outermost shells only provided they remain in contact.

In either case, electron transfer or electron sharing, it takes considerable energy to pull the atoms apart, and under ordinary circumstances, they remain together. Such atom combinations are called *molecules*, from a Latin word meaning "little object."

Sometimes two atoms in contact are enough to produce stability. Two hydrogen atoms form a hydrogen molecule; two nitrogen atoms, a nitrogen molecule; and two oxygen atoms, an oxygen molecule.

Sometimes it takes more than two atoms in contact to fill all the shells. The water molecule is made up of one oxygen atom and two hydrogen atoms; the methane molecule is made up of one carbon atom and four hydrogen atoms; the carbon dioxide molecule is made up of one carbon atom and two oxygen atoms; and so on.

In some cases millions of atoms can form a molecule. This is because carbon atoms in particular can each share electrons with each of four other atoms. Long chains and complicated rings of carbon atoms can therefore form. Such chains and rings form the basis of the molecules characteristic of living tissue. The molecules of proteins and nucleic acids in the human body and in all other living things are examples of such *mac-*

romolecules (*macro* is from a Greek word meaning "large").

Atom combinations in which electrons are transferred can bring about the formation of *crystals* in which the atoms exist in the countless millions all lined up in even rows.

On the whole, the larger a molecule and the less evenly the electric charge is distributed over it, the more likely it is that many molecules will cling together and that the substance will be liquid or solid.

All the solid substances we see are held together strongly by the electromagnetic interactions that exist first between electrons and protons, then between different atoms, and then between different molecules.

What's more, this ability of the electromagnetic force to hold myriads of particles together extends outward indefinitely. The nuclear interaction, which involves an attraction that fades exceedingly rapidly as distance is increased, can produce only the tiny atomic nucleus. The electromagnetic force, which fades only slowly with distance, can cluster everything from dust particles to mountains; it can produce a body the size of the Earth itself and bodies far larger still.

The electromagnetic force is intimately concerned with us in more ways than merely making it possible for us and for the planet we live on to be held together. Every chemical change is the result of shifts or transfers of electrons from one atom to another. This includes the very delicate and versatile shifts and transfers in the tissues of living beings such as us. All the changes that go on within us—the digestion of food, the contraction of muscles, the growth of new tissue, the sparking of nerve impulses, the generation of thought within the brain, all of it—is the result of changes under the control of the electromagnetic force.

Some of the electron shifts liberate considerable energy. The energy of a bonfire, of burning coal or oil, as well as the energy produced within living tissue, is the result of changes under the control of the electromagnetic force.

DENSITY

As the atoms or molecules of a given piece of matter move farther apart because of rising temperature or for any other reason, there comes to be less mass in a particular fixed volume of that matter. The reverse is true if the atoms or molecules come closer together.

The quantity of mass in a given volume is referred to as *density*; so what we are saying is that when matter expands, its density decreases, and when matter contracts, its density increases.

Scientists, using the metric system, measure mass in *grams* and volume in *cubic centimeters*. A gram is a rather small unit of mass, only about 1/28 ounce or 1/450 pound. As for a cubic centimeter, it is equal to about 1/16 cubic inch.

To give you a typical density, one cubic centimeter of water has a mass of one gram. (This isn't a coincidence. The two measures were originally chosen in the 1790s to fit together this way.) This means that we can say water has a density of 1 gram per cubic centimeter or, in abbreviated form, 1 g/cm^3.

Changes in density are not just a matter of expansion or contraction. Different substances have different densities because of the very nature of their structure.

Gases have densities much less than that of liquids, because gases are made up of separate atoms or molecules with little attraction for one another. Whereas liquid molecules are in virtual contact, the atoms or molecules of gases move about rapidly, bouncing off one another and in this way remaining far apart. Most of the volume of a gas is made up of the empty space between atoms or molecules.

For instance, a sample of hydrogen gas prepared on Earth at ordinary temperatures and pressures would have a density of roughly 0.00009 (or 9×10^{-5}) g/gm³. Liquid water is a little over 11,000 times as dense as hydrogen gas.

The density of the hydrogen could be made still lower if the hydrogen molecules (or separate atoms, for that matter) were allowed to move farther apart. In outer space, for instance, there could be so little matter that there is, on the average, only one atom of hydrogen in every cubic centimeter. In that case, the density of outer space would be something like 0.00000000000000000000000017 g/cm³—vanishingly small, indeed. Water is about 600 billion trillion times as dense as outer space.

Different gases are quite likely to differ in density. Under similar conditions the atoms or molecules making up the gases have just as much empty space between them. The density then depends on the mass of the individual atoms or molecules. If, of two gases, one is made up of molecules with three times the mass of those of the other, then the density of the first is three times that of the other.

For instance, a gas with a particularly massive molecule is uranium hexafluoride. Each molecule is made up of one uranium atom and six fluorine atoms, and the whole is 176 times as massive as the hydrogen molecule, with its two hydrogen atoms. Uranium hexafluoride is a liquid that turns into gas with gentle heating, and the density of that gas is about 0.016 g/cm^3. Liquid water is only 62.5 times as dense as that gas.

Still, any gas, even uranium hexafluoride, is mostly empty space. If such a gas is compressed —if, for instance, it is put into an airtight container the walls of which are then forced together—the gas molecules are pushed closer to one another, and the density increases.

The same effect is produced even more efficiently if the temperature is lowered. The gas molecules come closer together, and at some low-enough temperature the gas becomes a liquid, where the molecules are in virtual contact.

If hydrogen is cooled to very low temperatures, it not only liquefies, but at 14 degrees above absolute zero, it freezes. The molecules are not only in contact, but they remain more or less fixed in place, so that the substance is now a solid.

Solid hydrogen is the least dense solid in existence, with a density of 0.09 g/cm^3, and is only a tenth as dense as solid water. However, solid hydrogen, with its low density, is still over five times as dense as the very dense gas uranium hexafluoride.

In general, the density of liquids and solids also increases as the mass of the individual atoms and molecules of which they are composed increases. A solid made up of massive atoms is usually denser than one made up of less massive atoms.

Usually—however, not invariably. Here the situation is more complex than in the case of gases.

The comparative mass of different atoms is given by a figure known as the *atomic weight*. The atomic weight of hydrogen is approximately 1, so the atomic weight of any other atom gives you an approximate idea of the number of times heavier than a hydrogen atom it is. For instance, the aluminum atom has an atomic weight of about 27, while the iron atom has one of about 56. The iron atom has 56 times the mass of a hydrogen atom and just a little over twice the mass of an aluminum atom.

The density of iron, however, is about 7.85 g/cm^3, while that of aluminum is 2.7 g/cm^3. Iron is almost three times as dense as aluminum.

If iron is made up of atoms twice as massive as those of aluminum, why is iron three times as dense? Why not only twice as dense?

The answer is that other factors come into play. For example, how much room is taken up by the electrons of a particular atom and how compact the atom arrangement is. Atoms whose electrons belly far out from the central nucleus are less dense than you would expect from their mass, which is, after all, concentrated in the tiny nucleus. The electrons represent almost empty space and if they spread out and take up more room, the density is lowered.

Thus, cesium, with an atomic weight of 132.91, has a density of only 1.873 g/cm^3 because its electrons take up a great deal of room. The much more compact atoms of copper, with an atomic weight of 63.54, less than half that of cesium, give copper a density of 8.95 g/cm^3, nearly five times that of cesium.

If, then, you want to know the substance with the greatest known density, you must look among the more massive atoms but not necessarily among the most massive of all. The naturally occurring element with the most massive atoms is uranium, with an atomic weight of 238.07. Its density is a high 18.68 g/cm^3, twice that of copper, but it does not set a record. There are no less than four elements with a density greater than that. They, along with uranium, are listed in Table 3 in order of increasing density.

TABLE 3—Elements of High Densities

Element	Atomic Weight	Density (g/cm^3)
Uranium	238.07	18.68
Gold	197.0	19.32
Platinum	195.09	21.37
Iridium	192.2	22.42
Osmium	190.2	22.48

The rare metal osmium holds the record. Of the materials making up the Earth's crust or which can be obtained from it, this is the densest. Imagine an ingot of pure osmium the length and width of a dollar bill and 2.54 centimeters (1 inch) thick. That's not a large ingot, but it would weigh 5.85 kilograms (a kilogram is equal to a thousand grams), or nearly 13 pounds.

GRAVITATION

So far in this book we have talked quite a bit about the nuclear force and the electromagnetic force, and we have dismissed the weak force as comparatively unimportant for our purposes here. We have barely mentioned the gravitational force, however, and that, as it happens, is the most important of all as far as this book is concerned. In fact, we will speak of it so often that we may as well save syllables by referring to the gravitational force simply as *gravitation*, when that seems natural.

Gravitation affects any particle with mass, hadrons, leptons, and any combination of these—which means all the objects we see around us on Earth and in the sky.* We can expand Table 2 now into Table 4 by adding the weak force and gravitation.

TABLE 4—Particles and the Four Forces

	Proton	Neutron	Electron
Nuclear force	Yes	Yes	No
Electromagnetic force	Yes	No	Yes
Weak force	No	No	Yes
Gravitational force	Yes	Yes	Yes

Of all four forces gravitation is weakest by far, as was indicated in Table 1. We can demonstrate this, rather than merely state it, by engaging in some simple mathematics.

Suppose we consider two objects with mass alone in the universe. The gravitational force between them can be expressed by an equation first

---

* There are some particles without mass that are not affected in the ordinary sense by gravitation. The particles of light and similar radiations, called *photons*, from a Greek word meaning "light," are massless for instance. So are certain uncharged particles called *neutrinos*. Both of these will crop up later in the book.

worked out in 1687 by the English scientist
Isaac Newton (1642-1727), and that is:

$$F(g) = \frac{Gmm'}{d^2} \quad \text{(Equation 1)}$$

In this equation $F(g)$ is the intensity of the gravitational force between two bodies, m is the mass
of one body, m' the mass of the other, d the distance between them, and G is a universal *gravitational constant.*

We must be careful about our units of measurement. It is customary to measure mass in grams
and distance in centimeters. (A centimeter is
equal to just about 2/5 inch.) G is measured in
somewhat more complicated units that need not
concern us here. If we use grams and centimeters,
we will end up by determining the gravitational
force in units called *dynes*.

The value of G is fixed, as far as we know,
everywhere in the universe.* Its value in the units
we are using for it is 6.67×10^{-8}, or 0.0000000667.
Let's suppose that the two bodies in question are
exactly 1 centimeter apart, so that $d = 1$ and
therefore $d^2 = d \times d = 1 \times 1 = 1$. Equation 1
therefore becomes in this case:

$$F(g) = 6.67 \times 10^{-8} mm'. \quad \text{(Equation 2)}$$

Suppose now that we are dealing with an electron and a proton. The mass of the electron (m)
is 9.1×10^{-28} grams. The mass of the proton (m')
is 1.7×10^{-24} grams. If we multiply these two
figures and multiply the product by 6.67×10^{-8},

---

* There is some question about this, a point which will come up
later in the book.

we end up with a final product of 1×10^{-58} dynes, or 0.00000000000000000000000000000000000000-0000000000000000000001 dynes (Here is an example of why exponential figures are used by scientists in preference to ordinary decimals.)

We can therefore say that for a proton and an electron separated by 1 centimeter the gravitational attraction between them can be represented as:

$$F(g) = 1 \times 10^{-58} \text{ dynes.} \qquad \text{(Equation 3)}$$

Next let's move on to the electromagnetic force and set up an equation for its intensity between two charged objects alone in the universe.

Exactly one hundred years after Newton worked out the equation for gravitational force the French physicist Charles Augustin de Coulomb (1736-1806) was able to show that a very similar equation could be used to determine the intensity of the electromagnetic force. The equation is:

$$F(e) = \frac{qq'}{d^2}. \qquad \text{(Equation 4)}$$

In this equation $F(e)$ is the intensity of the electromagnetic force between the two bodies, q is the electric charge of one body, q' the electric charge of the other, and d is the distance between them. Once again, distance is measured in centimeters, and if we measure the electric charge in what are called *electrostatic units*, it is not necessary to insert a term analogous to the gravitational constant provided the objects are separated by a vacuum. (Of course, since I am assuming the objects are alone in the universe, there is necessarily a vacuum between them.) Furthermore, if we use these units, $F(e)$ will also come out in dynes.

If, once again, we assume that the two objects in question are separated by 1 centimeter, d^2 is again equal to 1 and the equation becomes:

$$F(e) = qq'. \qquad \text{(Equation 5)}$$

Suppose that we are still dealing with an electron and a proton. The two particles have equal electric charges (even though they are opposite in sign), each one being 4.8×10^{-10} electrostatic units. The product qq' is equal to $4.8 \times 10^{-10} \times 4.8 \times 10^{-10} = 2.3 \times 10^{-19}$ dynes.

Therefore for an electron and proton which are 1 centimeter apart, the electromagnetic force between them is:

$$F(e) = 2.3 \times 10^{-19} \text{ dynes.} \qquad \text{(Equation 6)}$$

If we want to find out how much stronger the electromagnetic force is than the gravitational force, we must divide $F(e)$ by $F(g)$. Since both intensities are measured in dynes under the conditions I've set up, the dynes will cancel, and we'll end up with a "pure" number, a number without units.

If we divide Equation 6 by Equation 3, we have:

$$\frac{F(e)}{F(g)} = \frac{2.3 \times 10^{-19}}{1 \times 10^{-58}} = 2.3 \times 10^{39}. \qquad \text{(Equation 7)}$$

In other words the electromagnetic force is 2,300,000,000,000,000,000,000,000,000,000,000,-000,000 times as strong as the gravitational force.

To get an idea of how enormous this difference

in intensity is, suppose we consider the gravitational force to be represented by a mass of one gram. What mass would we then have to use to represent the electromagnetic force? It would have to be a mass equal to a million bodies the mass of our Sun.

Again, suppose that the intensity of the gravitational force is symbolized by a distance equal to the width of one atom. The intensity of the electromagnetic force would then have to be symbolized by a distance a thousand times the width of the entire known universe.

Gravitation, then, is by far the weakest of the four forces. Even the so-called weak force is 10,000 trillion trillion times as strong as gravitation.

It is no wonder, then, that nuclear physicists, when studying the behavior of subatomic particles, take into account the nuclear force, the electromagnetic force, and the weak force but totally ignore gravitation. Gravitation is so weak that it simply never influences the course of events within atoms and atomic nuclei by a measurable amount.

This is also the case in chemistry. In all the considerations of the various chemical changes in the body and in the nonliving environment outside, only the electromagnetic force need be considered—with some interest in the nuclear force and the weak force in the case of radioactivity—but never gravitation. Gravitation is so weak that it introduces no measurable effect on ordinary chemical changes.

Then, why should we worry about gravitation at all?

Because somehow it's there and because, despite its incredible weakness, it somehow makes itself felt. We realize that every time we fall down. We know that if we fall as small a distance as that from a third-story window to the ground, we are very likely to be killed by the pull of gravitation. We know that it is gravitation that holds the Moon in orbit about the Earth and the Earth in orbit about the Sun. How is this possible for so weak a force?

Let's consider the four forces again. The nuclear force and the weak force decrease so rapidly with distance that they need not be considered outside such objects as atomic nuclei.

The electromagnetic force and the gravitational force, however, both decrease only as the square of the distance, and this is a slow enough rate of decrease to make it possible for both forces to make themselves felt at great distances.

There is this crucial difference between the two forces, however. There are two opposing kinds of electric charge and, as far as we know, only one kind of mass.

In the case of the electromagnetic force, there are attractions (between unlike charges) and repulsions (between like charges). Since the electromagnetic force is so strong, the powerful repulsion between like charges tends to scatter them, preventing any buildup of a sizable number in any one place. The equally powerful attraction between unlike charges tends to pull these together quite well, neutralizing the charges. In the end, the positive and negative charges (which are present in the universe in equal quantities, as far as we know) are thoroughly intermixed, and nowhere is there anything more than the tiniest excess of either charge.

Therefore, while the electromagnetic interaction is powerful and overwhelming in holding electrons in the neighborhood of the nucleus and in holding neighboring atoms together, one sizable chunk of matter has very little electromagnetic attraction or repulsion for another sizable chunk of matter some distance away, since in both chunks of matter the two different kinds of charge are so well mixed that both chunks end up having just about zero overall charge.*

Since there is only one kind of mass, however, there is only gravitational attraction. As far as we know, there is no such thing as gravitation repulsion. Every object with mass attracts every other object with mass, and the total gravitational force between any two bodies is proportional to the total mass of the two bodies taken together. There is no upper limit. The more massive the bodies, the more gravitational force acting between them.

Consider an object like the Earth, which has a mass equal to 3.5×10^{51} times that of a proton. In other words, it is 3,500 trillion trillion trillion trillion times as massive as a proton. Therefore the Earth produces a gravitational field that is 3.5×10^{51} times that of a single proton. Another way of looking at it is that every particle in the Earth that possesses mass—every proton, neutron, and electron—is the source of a tiny gravitational field,

---

* It is possible to remove some electrons from an object by friction, leaving it with a small positive charge, or to add some electrons, leaving it with a small negative charge. Such bodies can attract or repel each other and other objects, but the force involved is inconceivably tiny compared to what it would be if all the charged particles in either body could exert their full electromagnetic force.

and all of them melt together and add up to the overall gravitational field of the Earth.

The Earth also has electromagnetic fields, with every proton and electron in it acting as a source. The proton fields and the electron fields tend to cancel, however, so that the overall magnetic field of the Earth is very small indeed. It is enough to pull at the needle of a compass and to deflect charged particles coming from the Sun and elsewhere, but it is terribly weak for an object the enormous size of the Earth made up of so many charged particles.

Therefore, even though the gravitational force is so much weaker than the electromagnetic force when single particles are being considered, the gravitational force of the Earth as a whole is far greater than its electromagnetic force. The gravitational force of the Earth is strong enough to make you feel it unmistakably, and to kill you if you are not careful.

The enormous gravitational field of the Earth is capable of interacting with the lesser field of the Moon so that the two bodies are held strongly together. Gravitational forces hold the planets and the Sun together. There are measurable gravitational forces between the planets and between different stars.

Indeed, it is the gravitational force and *only* the gravitational force that holds the universe together and dictates the motion of all its bodies. All the other forces are localized. Only the gravitational force, by far the weakest of all, through the combination of being long range and of displaying attraction only, guides the destinies of the universe.

In particular, it is the gravitational force that is the key to any consideration of black holes, so you see we are on the main highway to them now and must study the landmarks as we proceed.

2 THE PLANETS

THE EARTH

ONE EARLY LANDMARK en route to the black hole (though it was never dreamed of as such at the time) was the determination of the mass of the Earth, something which was carried through by way of the gravitational force.

Newton had determined that the intensity of the gravitational field produced by any object is proportional to its mass. Indeed, that is another way of defining mass: that property of matter that produces a gravitational field.

This is not how I defined mass earlier in the book. Then I described it as that property of matter that makes it necessary to use a force of some sort to produce a change in motion of the matter, either in speed or in direction. The greater the force necessary to produce a certain change in the motion, the greater the mass of the body to which the force is applied.

The first definition of mass given just above is sometimes called the *gravitational mass*. The second, because it involves the reluctance of matter to undergo a change in its motion, a property called *inertia*, is referred to as *inertial mass*. Gravitation and inertia seem to be two entirely different properties, and there seems no reason to suppose that the two kinds of mass should match each other exactly, that whenever one mass has twice the inertia of another, it also has a gravitational field of twice the intensity. Nevertheless, that's the way it seems to work out. No one has ever been able to show any distinction between gravitational mass and inertial mass, so it is now taken for granted that the two are identical.

Thus, the gravitational field of the Earth exerts a force on a falling body so that it undergoes a change of motion, or *acceleration*, and falls faster and faster. Since inertial mass and gravitational mass are the same, we can suppose that the increase in the rate of speed with which an object falls can be used to measure the intensity of Earth's gravitation.

This acceleration was first measured back in the 1590s by the Italian scientist Galileo Galilei (1564-1642). It is equal to 980 centimeters per second per second This means that each second a falling body is moving 980 centimeters per second faster than it was falling the second before.

Now let us go back to Newton's equation:

$$F = \frac{Gmm'}{d^2}. \qquad \text{(Equation 8)}$$

where F is the intensity of the gravitational field

and therefore the value of the acceleration of a
falling body, which, as I say, has long been known.
G is the gravitational constant, m is the mass of
the falling body, m' is the mass of the Earth, and
d is the distance between the falling body and the
Earth. What we're really interested in is the mass
of the Earth, so let's alter the equation by the
usual algebraic techniques to put the m' all by
itself on the left-hand side of the equation. The
equation becomes:

$$m' = \frac{Fd^2}{Gm}. \qquad \text{(Equation 9)}$$

If we have values for every symbol on the right-
hand side of the equation, we can multiply the
value of F by the value of d, multiply the product
again by d, divide this result by G, divide the quo-
tient by m, and that will give the value for m',
the mass of the Earth.

Well, that looks great, because we do have the
value of F to begin with, as I have just explained.
We also have the value of m, the mass of the falling body, because we can just weigh it on a balance to find the mass in grams.

The distance between the falling body and the
Earth is a little complicated. Newton showed that
when a body produces a gravitational field, that
field behaves as though it is produced by all the
mass of the body concentrated at its center of
gravity. When the body has a shape and properties that fulfill certain conditions of symmetry, the
center of gravity is at the geometric center of the
body. These conditions of symmetry hold for
the Earth and for all the sizable bodies that we
know of in the universe.

This means that the Earth acts as though its

gravitational field originates at its center; d therefore stands for the distance of the falling body from the center of the Earth, not from the Earth's surface. If the falling body is near the surface of the Earth, then the distance is equal to the radius of the Earth's sphere.

This value was first determined about 240 B.C. by a Greek geographer named Eratosthenes (276-192 B.C.). He determined the size of the Earth's sphere from the rate at which its surface curves, which he in turn determined by measuring the angle the rays of the Sun made to different parts of that surface at the same time. The radius (the distance from the surface of the Earth to its center) is equal to 637,000,000 centimeters.

Now we have the values of F, m, and d, but as late as the 1700s, we *didn't* have the value of G, and until we got the value of G, we couldn't use Equation 9 to calculate m', the mass of Earth.

Is there any way we can determine the value of G?

Well, if G is truly universal, then suppose we measure the intensity of the gravitational field between two lead balls and make use of another form of Equation 8. Algebraic techniques can convert it into:

$$G = \frac{Fd^2}{mm'} \qquad \text{(Equation 10)}$$

We can easily measure the mass of each of the lead balls, and that gives us m and m'. We can also measure the distance between them, and that gives us d. If we can then also measure the gravitational force between them and get F, we can solve the equation for G at once. Then we can put the value of G into Equation 9 and instantly calculate the mass of the Earth.

There is another catch. Gravitational forces are so incredibly weak in relation to mass that it takes a hugely massive object like the Earth to have a gravitational field intense enough to measure easily. Before we can work with objects small enough to deal with in the laboratory, we must have some device that can measure very tiny forces.

The necessary refinement in measurement came about with the invention in 1777 of the *torsion balance* by Coulomb (who worked out Equation 4, which I mentioned earlier in the book). In the torsion balance tiny forces are measured by having them twist a fine string or wire. To detect the twist, one need attach to the vertical wire a long horizontal rod balanced at the center. Even a tiny, almost imperceptible, twist would produce a measurable movement at the end of the rod. If the wire being twisted is thin enough and the attached rod is long enough, we can measure the twist produced by the tiny, tiny gravitational fields of ordinary-sized objects.

The wire or thread, you see, is elastic, so there is a force within it that tends to untwist it. The more it is twisted, the stronger the force to untwist it becomes. Eventually the untwisting force balances the twisting force, and the rod remains in a new equilibrium position. It is by measuring the extent to which the rod has twisted to achieve a new equilibrium that one can determine the intensity of force acting upon it.

In 1798 the English chemist Henry Cavendish (1731-1810) attempted the following experiment: He began with a rod 180 centimeters long and placed on each end a 5-centimeter-in-diameter lead ball. He next suspended the rod from its center by a fine wire.

Cavendish then suspended a lead ball a little over 20 centimeters in diameter on one side of one of the small lead balls at the end of the horizontal rod. He suspended another such ball on the opposite side of the other small lead ball. The gravitational field of the large lead balls would now serve to attract the small lead balls and twist the wire to a new position. From the change represented by the new position compared with the old Cavendish could measure the tiny gravitational attraction between the lead balls.

(Naturally, Cavendish enclosed the whole thing in a box and took every precaution to avoid having the wire stirred by air currents.)

Cavendish repeated the experiment over and over again until he was satisfied he had a good measurement of F. Since there was no problem in measuring the mass of the lead balls or the distances between the large balls and the small ones, he already had m, m' and d. He could now solve for G in Equation 10 and did so.

Using refinements of Cavendish's experiments, we now believe the mass of the Earth to be 5.983 $\times$ 10^{27} grams, or roughly 6,000 trillion trillion grams.

We can determine the density of any object by dividing its mass by its volume. The volume of the Earth had been worked out correctly, or nearly so, from Eratosthenes's figure for the Earth's circumference. Once Cavendish had worked out the mass of the Earth, it was therefore possible to determine the Earth's average density at once. It turned out to be 5.52 g/cm^3.

THE OTHER PLANETS

The importance of determining the mass of the Earth lies not only in itself but in the fact that it has made it possible for astronomers to determine the mass of large numbers of other objects in the universe.

There is the Moon, for instance, Earth's one satellite, which is 384,000 kilometers from us (a kilometer is equal to about five eighths of a mile) and which circles Earth once every 27 1/3 days.

More precisely, both Earth and Moon circle a common center of gravity. The laws of mechanics require that the distance of each body from that center of gravity be related to its mass. In other words, if the Moon were one half as massive as the Earth it would be two times as far from the center of gravity as the Earth is; if it were one third as massive as the Earth, it would be three times as far; and so on.

The position of the center of gravity of the Earth-Moon system can be determined by astronomers, and it turns out to be located about 1,650 kilometers under the surface of the Earth and about 4,720 kilometers from the center of the Earth. (It is the center that counts in gravitational affairs, remember.) The Moon goes around that point, and so does the Earth; Earth's center wobbles about it every 27 1/3 days.

The center of gravity is 81.3 times as far from the center of the Moon as it is from the center of the Earth, so the Moon has 1/81.3 or 0.0123 times, the mass of the Earth. The mass of the Moon is thus 7.36×10^{25} grams, but it is easier to express the value as a fraction of the Earth's mass.

Astronomers can go on to determine the mass of the other planets of the solar system relative to that of Earth. One way is by comparing the effect of a planet on its satellite with that of the Earth on the Moon.

The time in which a small satellite completes its orbit about its planet depends on two things only: the satellite's distance from the planet's center and the intensity of the planet's gravitational field.

For instance, the planet Jupiter has a satellite, Io, which is at nearly exactly the same distance from Jupiter that the Moon is from Earth. Yet Io circles Jupiter in 1 3/4 days, while the Moon circles Earth in 27 1/3 days.

It can be calculated that Jupiter's gravitation must be 318.4 times as intense as Earth's in order for Jupiter to be able to whip Io so quickly about itself. In other words, Jupiter must have a mass 318.4 times that of the Earth. Using this satellite method and others, one can determine the mass of all the sizable objects in the solar system.

In Table 5 the masses and densities of the nine planets of the solar system, and of our Moon as well, are given in order of distance from the Sun.

The intensity of the gravitational field of each of these bodies is in proportion to their mass, and as you can see, Earth has by no means the greatest gravitational intensity or the greatest mass among the planets of the solar system. There are four planets more massive than the Earth— Jupiter, Saturn, Uranus, and Neptune. Jupiter is the giant of the planetary system; it is about 2.5

TABLE 5—Mass and Density of the Planets

	Mass (Earth = 1)	Density (g/cm³)
Mercury	0.055	5.4
Venus	0.815	5.2
Earth	1	5.52
Moon	0.0123	3.3
Mars	0.108	3.96
Jupiter	317.9	1.34
Saturn	95.2	0.71
Uranus	14.6	1.27
Neptune	17.2	1.7
Pluto	0.1	4

times as massive as the other eight planets put together.

The intensity of the gravitational field of each planet (or of any body) decreases as the square of the distance, which means that the *relative* intensity of the gravitational field of two bodies of different mass remains the same at any distance.

For instance, a spaceship a million kilometers from Jupiter's center would feel Jupiter's gravitational pull to be 317.9 times as strong as it would feel Earth's gravitational pull to be if it were a million kilometers from Earth's center.

If the spaceship were to increase its distance from Jupiter's center from 1 million kilometers to 2 million kilometers, Jupiter's gravitational field would be only one fourth as intense at the new position as at the old. If the same thing were done in connection with Earth, Earth's gravitational field would also be only one fourth as intense at the new position as at the old. Jupiter's field at its new point would remain 317.9 times as strong as Earth's field at its new point.

Jupiter's gravitational field would be 317.9 times as strong as Earth's at every pair of corresponding points—but what if the points do not correspond?

There is one important occasion when we would be forced to remain at a different distance from one planet's center than from another's. That is when we are standing on the surface first of one planet, then of another, with the two planets being different sizes.

We can demonstrate this best by comparing Earth with the Moon, since men have stood on both worlds and have confirmed what theory predicts.

The Earth's mass is 81.3 times that of the Moon, and for positions at equal distances from the center of each body the intensity of Earth's gravitational field is always 81.3 times that of the Moon's.

Suppose we are standing on the Moon's surface, though. We are then 1,738 kilometers from the Moon's center. If we are standing on the Earth's surface, we are 6,371 kilometers from the Earth's center.

The gravitational intensity on the surface of a body is its *surface gravity* (an important concept in the story of black holes), and in calculating that, we must take into account the differences in distance from the center. The distance of Earth's surface from Earth's center is 3.666 times the distance of the Moon's surface from the Moon's center.

Gravitational intensity weakens as the square of the distance, so Earth's surface gravity is weakened as compared with the Moon's surface gravity by a factor equal to 3.666 × 3.666, or 13.44. We must therefore divide Earth's innate gravitational intensity of 81.3 (compared to the Moon's) by 13.44, and that gives us an answer of 6.05.

Thus, although Earth has a mass 81.3 times that of the Moon, its surface gravity is only 6.05 times that of the Moon. To put it another way, the Moon's surface gravity is about one sixth that of the Earth.

In similar fashion we can calculate the surface gravity for all the bodies of the solar system. The four giant planets offer a problem because what we see as a "surface" is actually the outer edge of their huge atmospheres, whose thickness we cannot easily judge. We cannot even be certain that there is a solid or liquid surface anywhere. If we pretend, however, that we can come to rest at the top of that cloud layer and calculate the intensity of the gravitational field at that point, we can call it the surface gravity. With that in mind, we can prepare Table 6.

TABLE 6—Surface Gravity

	Surface Gravity (Earth = 1)
Mercury	0.37
Venus	0.88
Earth	1.00
Moon	0.165
Mars	0.38
Jupiter	2.64
Saturn	1.15
Uranus	1.17
Neptune	1.18
Pluto	0.4

ESCAPE VELOCITY

It is the Earth's gravitational field that lies behind the old saying "Everything that goes up must come down." Any object hurled into the air at some particular velocity is under the constant pull of Earth's gravitation. It therefore loses velocity steadily until it comes to a momentary halt at some point above the Earth's surface. Then it begins to fall, gaining velocity steadily until it hits the ground at the same velocity with which it was was originally hurled upward.*

If one of two objects is hurled upward at a greater velocity than the other, it will take longer for it to lose its velocity; it therefore climbs higher before the turnaround. It might seem that no matter how great the velocity with which an object began its upward climb, that velocity would eventually be eroded away. It might climb a hundred kilometers, a thousand kilometers, but eventually the relentless pull of the gravitational field would have its way.

So it would seem—and so it would be if the intensity of the gravitational field did not weaken with distance.

Earth's surface gravity exerts a certain force on an object on the surface, which is 6,371 kilometers from Earth's center. The intensity of gravitation decreases as any object subject to that force rises up from the surface and increases its distance from Earth's center. The decrease in intensity is proportional to the square of the

---

* Actually air resistance complicates the situation and further slows the object both going up and coming down. We are going to pretend in this section, however, that air resistance doesn't exist. That involves only a small change and doesn't alter the essence of the argument.

distance—but distance from the center, not from the surface.

Suppose we rise into the stratosphere, some 35 kilometers above Earth's surface. This is a great height by ordinary standards, but the distance from the center of the Earth increases only from 6,371 kilometers to 6,406 kilometers. That is not much of a change; the gravitational intensity at this height is still 98.9 percent that on the surface itself. A human being weighing 70 kilograms on the surface would still weigh 69.23 kilograms in the stratosphere. In ordinary life, then, we are not conscious of any significant change in the intensity of Earth's gravitation, so we never allow for that change.

Suppose, however, that an object rises a really great distance, say to a height of 6,371 kilometers above the Earth's surface. It is then 6,371 + 6,371, or 12,742 kilometers from the Earth's center. Its distance from the center will have increased by a factor of two, and the gravitational intensity will have decreased to one fourth of what it was at the surface.

If we imagine an object hurled upward with such velocity that it reaches the stratosphere before that velocity is lost, then we see that in the later stages of its upward flight the gravitational intensity is slightly lower than it was in the earlier stages. The further loss of velocity is less, then, than would be expected if gravitation intensity remained the same all the way up. The object moves up somewhat higher than would be expected before that momentary halt and turnaround.

Next imagine that a second object is hurled upward with an initial velocity double that of the first object. By the time the second object has reached the height at which the first object had lost all its velocity, it will have lost only half its velocity. It would now be moving at the velocity the first object had had to begin with.

Will the second object now climb an additional distance equal to the total distance the first object had climbed?

No, for the second object is now making its additional climb through a region of somewhat weaker gravitation. It loses velocity more slowly and climbs through a greater distance than the first object did from the surface.

Because of the decline in gravitational intensity with height, doubling the initial velocity of an object hurled upward *more than doubles* the height it reaches. In Table 7 we see the height to which objects rise above the surface of the Earth at given initial velocities.

TABLE 7—Rising Bodies

Initial Velocity (km/sec)	Maximum height above Earth's surface (km)
1.6	130
3.2	560
4.8	1,450
6.4	3,100
8.0	6,700
9.6	17,900

As the initial velocity increases, the maximum height increases too, and it increases more and more rapidly as the object moves into regions of weaker and weaker gravitation. Between the first and last entries in the table the initial velocity has increased by a factor of 6, but the maximum height has increased by a factor of 140.

There comes a point where an object rises so rapidly that its velocity decrease matches the decline in gravitational intensity. When it has lost half its velocity, the gravitational intensity has also sunk to half, so that now it would take as much time for that weakened intensity to remove the half velocity left than it would have taken the full gravitational intensity to remove the full velocity. The object moving upward continues to lose velocity but at an ever slower pace as gravitation grows weaker and weaker. The rising body never entirely loses all, and in that case, what goes up doesn't come down because it never quite stops going up.

The minimum velocity at which this happens is the *escape velocity*.

The escape velocity from Earth's surface is 11.23 kilometers per second. Anything hurled upward from Earth at a velocity of 11.23 kilometers per second or more will go up and never come down; it will move farther and farther from Earth. Anything moving upward with an initial velocity of less than 11.23 kilometers per second (with no further push added to what it already has *) will return to Earth.**

---

* An object that has an initial velocity and no added push is in *ballistic flight* and must begin with the escape velocity or more to move indefinitely away from the Earth. A rocket ship, however, can be continually pushed by its rocket exhaust so that, although it may move at less than escape velocity, it can get as far above Earth as desired. However, where living things are not involved, motion in the universe is almost always ballistic motion, with one initial impulse and no more.

** If an object is moving less than escape velocity but not less than about 70 percent of escape velocity, and if it also has a sideways motion, then it may not escape from Earth but may not drop back to the surface, either. It may take up an orbit around the Earth and remain in that

The value of the escape velocity depends on the intensity of the gravitational field. As that intensity declines, the escape velocity declines, too. It turns out that as we increase our distance from Earth's center, the escape velocity declines as the square root of that distance.

Suppose we are in space 57,400 kilometers from the Earth's center. That would place us nine times as far from the center as we would be if we were on Earth's surface. The square root of nine is three, and that means the escape velocity at the height of 57,400 kilometers from Earth's center is only one third of what it is at the surface of the Earth. At the height it is 11.23/3, or 3.74 kilometers per second.

The escape velocity is different for different worlds. A world that is less massive than Earth and has a lower surface gravity also has a lower escape velocity from its surface. The escape velocity from the Moon's surface, for instance, is only 2.40 kilometers per second.

On the other hand, worlds that are more massive than Earth have higher escape velocities than it has. In Table 8 the escape velocities from the various planets are given, as measured from the visible surface (meaning the upper edge of the cloud cover in the case of the giant planets).

It is not surprising that the giant of the planetary system, Jupiter, has the highest escape velocity.

What's more, because it is so voluminous, Jupiter has a gravitational field that declines with distance more slowly than Earth's does. Since

---

orbit indefinitely. An astronaut orbiting the Earth just a couple of hundred kilometers above the surface must travel at least 7.94 kilometers per second to remain in orbit.

TABLE 8—Escape Velocities from the Planets

	Escape Velocity (km/sec)
Mercury	4.2
Venus	10.3
Earth	11.23
Moon	2.40
Mars	5.0
Jupiter	60.5
Saturn	35.2
Uranus	21.7
Neptune	24
Pluto	5

Earth's surface is 6,371 kilometers from the center, its gravitation weakens to 1/4 its surface value at a height of 6,371 kilometers above the surface. At a height of 19,113 kilometers above the surface, the distance from the center of the Earth is 4 times what it was at the surface, and Earth's gravitation is only 1/16 its surface value.

Jupiter, however, has a surface that is 71,450 kilometers from its center. Therefore one must rise to a height of 71,450 kilometers above Jupiter's surface before its gravitation drops to 1/4 its surface value, and to a height of 214,350 kilometers above its surface before its gravitation drops to 1/16 of its surface value.

Jupiter's gravitational intensity drops so much more slowly than Earth's that at equal distances far out in space Jupiter's gravitational intensity is 317.9 times that of Earth (what it should be considering their comparative masses), even though

Jupiter's surface gravity is only 2.64 times that of the Earth.

Jupiter's escape velocity also decreases with distance more slowly than Earth's does. From Jupiter's surface the escape velocity is only 5.4 times that from Earth's surface. The escape velocity from Jupiter decreases so slowly with distance, however, that even at a height of 2 million kilometers above Jupiter's surface, it is still equal to that from Earth's surface.

PLANETARY DENSITY AND FORMATION

Despite the size of Jupiter's surface gravity and escape velocity as compared with that of Earth, the impression we ought to get is of surprise at Jupiter's feebleness.

Jupiter is after all more than three hundred times as massive as Earth and has a gravitational field more than three hundred times as intense as Earth's in consequence; yet the surface gravity of Jupiter is less than three times that of Earth, and its escape velocity is less than six times that of Earth. The same disparity between the gravitational intensity on the one hand and the surface gravity and escape velocity on the other can be seen in the other giant planets.

The reason for this is that the giant planets are so bulky that their surface (their cloud-layer surfaces, anyway) are anywhere from nearly four to over eleven times as far from their planetary centers as Earth's surface is from its center.

And that is not the whole explanation. The giant planets have low densities, which means that the matter within them is not compactly packed together but is spread out to take up a more than normal volume by Earth standards. Their surfaces are thus farther out than they would be if the giant planets were denser.

Suppose we indulge in a fantasy and imagine that the planet Saturn could somehow be pushed together, or compressed, to the point where its average density would be that of the Earth. It would have to be compressed to the point where its volume would be only one eighth of what it is now. Its radius would only be half what it is now: 30,000 kilometers instead of the present 60,000.

Saturn would still have all its mass under these conditions. Both its mass and the intensity of its gravitational field would still be 95.2 times that of Earth. The surface would still be farther from the center than is true of Earth, but not so much farther so the surface gravity, when Saturn is compressed to Earth density, would not be 1.15 times that of Earth, but 4.60 times.

Suppose we fantasize that Jupiter, too, could be compressed to the average density of Earth. Its volume would be only one fourth of what it is now, and its radius five eighths of what it now is: 44,200 kilometers instead of the present 71,400. With its mass intact and its surface that much closer to the center Jupiter's surface gravity would be just about 7 times Earth's surface gravity.

Is there any other way in which we can get closer to the center of a world and therefore increase the gravitational intensity? For instance, if we burrowed down into the crust of the Earth itself, would the gravitational force upon ourselves increase steadily as we approached the center? No!

Suppose we imagine the Earth had an even

density of 5.52 g/cm³ all the way through and that we could somehow burrow into it freely. As we dug down, part of the Earth's structure would be above us. In fact, a whole outer sphere of Earth's structure would be farther from the center than we were. Newton's mathematics showed that this outer part would not contribute to the gravitational force pulling us toward the center. Only the part of the Earth that was nearer to the center than we were at any particular time would contribute to that, and there would be less and less of it as we burrowed deeper and deeper.

This means that the gravitational pull upon us would grow weaker and weaker as we burrowed into the Earth until we reached the very center of the planet, when it would be zero. At the center of the Earth, or of any spherical world, all the mass of the world would be pulling at us in the direction away from the center because it would all be above us. It would, however, be pulling outward in all directions equally, and the pulls would cancel out, leaving us with zero gravity.

If we could imagine a sizable hole at the center of the Earth, or of any spherical world, there would be zero gravity at every point within the hole. Science-fiction stories have been written in which the Earth was imagined as hollow with an inhabited interior surface lit by a Sun-like object at the center. Edgar Rice Burroughs's stories about "Pellucidar" are an example. Any inhabitants of such a world would, however, feel no gravitational pull holding them to that interior surface, but would float about freely in the internal space—something Burroughs didn't realize.

No, the only way to increase the gravitational pull is to compress the entire world, packing *all* the mass more tightly together so that you can approach the center while keeping *all* the mass between you and the center—a concept that is of key importance in understanding the black hole.

The only thing in the universe that can so compress a world is gravitation itself, and it has done so in the past, when, for instance, the planets of our solar system were forming.

At the start, the material out of which the planets were formed was a vast mass of dust and gas. Most of this material was hydrogen, helium, carbon, neon, oxygen, and nitrogen, with hydrogen making up perhaps 90 percent of all the atoms. All of it, slowly swirling in separate turbulent whirlpools, slowly came together under the weak, but ever sustained pull of the mutual gravitation of all the atoms and molecules.

The more closely the material came together, the more it was compressed, the more the separate gravitational fields of the constituent parts overlapped and reinforced one another. The gravitational intensity increased, and the further compression took place faster—and faster.

Most of the material remained gaseous. The helium and neon remained as separate atoms. The hydrogen atoms combined into two-atom hydrogen molecules but remained as separate molecules. The carbon atoms each combined with four hydrogen atoms to form methane molecules, which remained separate. The nitrogen atoms each combined with three hydrogen atoms to form ammonia molecules, which remained separate. The oxygen atoms each combined with two hydrogen atoms to form water molecules, which remained separate.

There were two moderately common elements

that didn't remain as separate atoms or as separate small molecules. These were silicon and iron. Silicon atoms combined with oxygen atoms but, in the process, did not form molecules that remained separate. In this case, the electromagnetic force kept on working to pile more and more silicon-oxygen combinations together without limit. These combinations, called *silicates,* could grow to be dust particles, then pebbles, then rocks and boulders. Atoms of other elements that would fit into the silicate structure were added: magnesium, sodium, potassium, calcium, aluminum, and so on. It is this mixture of silicates that forms the rocky materials of the Earth's crust with which we are so familiar.

Iron atoms clung to one another for the most part, together with other metals, such as cobalt and nickel, that mixed with them freely.

Thus, as the dust and gas swirled inward ever more tightly, ever larger pieces of rock and metal (or combinations of both) formed. Since the metal was denser than the rock, it responded more to gravitational pull. As a world formed, the metal would sink toward the center, forming a core, while the rocky material remained in a shell outside the metal.

The Moon and Mars are built chiefly of rock. Mercury, Venus, and Earth are built of rock and metal. Tiny solid bits of matter still strew space, and some strike Earth's atmosphere as *meteors,* which, if they survive to reach Earth's solid or liquid surface, are called *meteorites.* Some meteorites are rock, some metal, and some a mixture.

Small worlds like the smaller asteroids are not large enough to have a gravitational field intense enough to hold them together. They are held together by the electromagnetic force within and between the atoms, which is, of course, enormously more intense than the gravitational force of such small bodies.

Atoms and molecules that remain separate and don't build up endless electromagnetically held combinations won't cling to worlds by electromagnetic interaction but can only be held gravitationally. The separate atoms and molecules that make up a gaseous atmosphere are examples of this.

Small worlds lack gravitational fields intense enough to hold such gases. The Moon, therefore, with a surface gravity only one sixth as strong as Earth's, cannot hold gas molecules and does not have an atmosphere. What's more, it cannot hold molecules of liquid that are *volatile*, that is, that evaporate and turn into gases easily. For that reason the Moon has no free water on its surface. Worlds even smaller than the Moon would also lack atmospheres and volatile liquids.

Mercury, with a surface gravity 2.3 times that of the Moon but still only three eighths that of the Earth, has neither atmosphere nor ocean, while Mars, with a surface gravity about like that of Mercury, manages to have a very thin atmosphere —about 0.006 times as dense as ours—together with traces of water.

Why?

The answer is that temperature has an effect. The higher the temperature, the more rapidly the atoms and molecules of gases move, the more likely it is that some of them will move at speeds greater than the escape velocity of the planet to which they belong, the more likely it is that the atmosphere (if any exists to begin with) will dissipate into space, and the less likely it is that the atmosphere will have formed in the first place. The lower the temperature, the less rapidly the atoms and molecules of gases move, the less likely it is that any of them will move at speeds above the escape velocity, the less likely it is that the atmosphere will dissipate, and the more likely it is that the atmosphere will form in the first place.

Mars has the same surface gravity as Mercury has, but Mars is nearly four times as far from the Sun as Mercury is and is therefore considerably cooler. Whereas Mercury's surface can reach temperatures of 350°C, the average Martian surface temperature is only 20°C.

Consider Titan, the largest satellite of the planet Saturn. Titan's surface gravity is probably not more than half that of Mars, but Titan has a surface temperature of about −180°C, only 90 degrees above absolute zero. It therefore possesses an atmosphere that seems to be denser than that of Mars and may be almost as dense as that of Earth.

The less massive an atom or molecule, the more quickly it moves at a given temperature, the more likely it is that it will escape into space, and the more difficult it is to hold onto as part of an atmosphere.

Thus, the Earth's gravitational field is intense enough to hold argon atoms (with an atomic weight of 40). It can also hold carbon dioxide, since the carbon atom it contains has an atomic weight of 12 and the two oxygen atoms it contains have a total atomic weight of 32, making an overall *molecular weight* of 44.

In the same way Earth's gravitational field is intense enough to hold oxygen (molecular weight

32) and nitrogen (molecular weight 28), but not helium (atomic weight 4) or hydrogen (molecular weight 2).

If the gradual buildup of material forming a planet becomes large enough to give rise to a gravitational field intense enough to hold even helium or hydrogen, the planet then starts to grow rapidly, since helium and hydrogen are the most common of the starting materials. The planet, in effect, snowballs, since the further it grows, the more intense its gravitational field and the more effectively it can continue to gather more helium and hydrogen.

This happens more easily farther from the Sun, where it is cooler and the light gases are made up of atoms and molecules that are moving comparatively slowly. The result is the formation of the giant planets Jupiter, Saturn, Uranus, and Neptune, relatively far from the Sun. It is because they are made up largely of the light elements that they possess such low densities.

Planets forming closer to the Sun, where temperatures are higher, cannot hold onto the light elements; they are built up chiefly or entirely of those less common atoms that can hold together by electromagnetic force. Hence it is the smaller planets of rock and metal, with high densities, that make up the inner solar system.

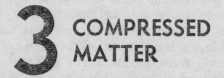

3 COMPRESSED MATTER

PLANETARY INTERIORS

As THE PARTICLES making up a planet come together—growing to pebbles, boulders, mountains, and worlds—they heat up. Gravitation produces an acceleration motion inward; the larger the growing fragments become and the faster they move, the more *kinetic energy* (*kinetic* is from a Greek word meaning "motion") they possess. The larger fragments, *planetesimals*, which bang into the growing world have the energy to gouge out huge craters. These are eliminated by the crashes and the ever more intense crater gouging that follow, until finally the last few remain indefinitely.

We see the craters that mark the last collisions on the Moon, on Mercury, on Mars, and on the two small Martian satellites, Phobos and Deimos. We could surely see them on Venus if we could get a good look under the clouds and on Jupiter's satellites if we could get pictures of them in sufficient detail.

Undoubtedly the Earth had its share of craters, too. On the Earth, however, running water and the action of living things have eroded them, and only faint traces can be seen.

All the kinetic energy of the crashing together of rapidly moving bodies is not lost. Energy cannot be lost; it can only be changed into other forms. In this case the kinetic energy is turned into heat and is concentrated at the center of the world that is formed. This is true of Earth and, undoubtedly, of all worlds large enough to have received a great deal of kinetic energy in the process of formation. This internal heat is the product, in the last analysis, of the energy of the gravitational field as it is concentrated more and more intensely in the process of planet making.

In the case of the Earth evidence was early gained that the interior is hot. When one digs deep mines into the Earth, the temperature goes up steadily as one probes deeper. There are also indications of internal heat in the form of hot springs and volcanoes (which probably gave ancient man the idea of a fiery hell underground).

Modern knowledge of Earth's interior stems from the analysis of earthquake waves, which travel through the body of the planet. From the paths they take, the time it takes them to travel, and the manner in which they make or don't make sudden changes of direction a great deal can be inferred concerning the properties of the Earth's interior. The temperature is believed to rise steadily all the way to the Earth's center, and at the center the temperature may be as high as 5,000°C (nearly as hot as the 6,000°C of the Sun's surface).

The fact that the interior of the Earth is blazing hot means that much of its internal structure was (and still is) in the liquid state after it was formed and after the planet reached something like its present size. That means that if it were made up of different kinds of matter that do not readily mix with one another, they would separate, the denser varieties moving closer to the center and the less dense varieties floating on top of the denser ones.

This, indeed, happened. The Earth is chiefly made up of rocky silicates and a metal mixture of iron and nickel in a ratio of about nine to one. The metal settled in the center, where it now forms a *nickel-iron core*. Around it is the silicate *mantle*. The mantle is solid, for its temperature at its hottest (which is, of course, at its deepest point) is probably no more than 2,700°C, which is not enough to melt the rock. The core, with a considerably higher temperature, is hot enough to melt the iron; thus the Earth has a liquid core.

The heat in the Earth's interior was originally formed in the early stages of the planet's history— 4,600,000,000 years ago. (A billion years is sometimes called an *eon*, so that we may say the Earth was formed 4.6 eons ago.) Perhaps by 4 eons ago the major planetesimal collisions were over, and very little in the way of more kinetic energy was added to the Earth. Gravitation had completed its work of formation.

It would seem that in the 4 eons that have since passed, the internal heat should have leaked out of the Earth, and the whole planet should have cooled down. The rock of the mantle and crust is, to be sure, a very poor conductor of heat, so the internal heat could leak out only very slowly indeed, but 4 eons is a long, long time.

Actually, though, the Earth has as part of its constituents small quantities of elements like uranium and thorium that by means of the nuclear force and the weak force slowly break down over the eons and liberate heat. (After 4.6 eons of existence on Earth half the original uranium and four fifths of the original thorium still exists intact.) The heat liberated by these radioactive elements is not very much, but it adds up over the eons; it is at least as great as the amount of internal heat leaking out. What was begun by the gravitational force is now maintained by the nuclear and weak forces; Earth's interior will therefore not cool down for many eons to come.

Naturally a planet that is larger than the Earth has received much more kinetic energy in the process of formation. In the first place up to hundreds of times more mass has come crashing into the growing planet. Then, too, because of the steadily more intense gravitational field those masses struck at greater speeds. Both mass and speed contribute to kinetic energy. A large planet would therefore have a hotter interior than Earth has (and a small planet would have a cooler one).

Consider Jupiter. In 1974 and 1975 two probes, Pioneer 10 and Pioneer 11, passed quite close to the planet (within 100,000 kilometers of its surface), and from the data received, scientists were able to estimate the interior temperatures of the vast planet.

The distance from the outer cloud layer of Jupiter to the center is 71,400 kilometers. By the time a depth of 2,900 kilometers below the cloud surface is reached (only 4 percent of the way to the center), the temperature is already some 10,000°C, twice as high as Earth's central point.

At a depth of 24,000 kilometers below the cloud surface, a third of the way to Jupiter's center, the temperature is 20,000°C. At the center itself the temperature has reached a whopping 54,000°C, nine times that of the surface of the Sun.

But it isn't only high temperature that is produced in planetary interiors by the gravitational interaction. High pressures are also produced.

Under the action of the gravitational field the outermost layers of a planet are pulled toward the center and push against the layers beneath, which are also pulled toward the center and push against the layers beneath them. This series of pushes is carried on all the way to the center, each deeper layer transmitting the push of everything above it and adding its own, so that the pressure goes steadily higher as one penetrates deeper and deeper into a planet.

A pressure is often measured as a certain weight distributed over a certain area—the number of grams pushing down on a square centimeter, for instance. Consider our atmosphere. It is pulled down against the surface of the Earth by gravitation with sufficient intensity to cause it to push against that surface with considerable pressure.

Every square centimeter of the Earth's surface receives the push (or the *weight*, which is what the push is often called) of 1,033.2 grams of air. We can say, then, that air pressure at sea level is 1,033.2 g/cm², which we can call 1 *atmosphere*. This pressure is also exerted on our bodies, but in every direction, both inside and out, so it cancels out and we are not aware of it.

The pressure of water in the ocean depths is much higher than that of air, since water is much denser than air and there is a greater mass of it to be pulled downward. At the deepest part of the ocean the water pressure is just over 1,000,000 g/cm^2, or about 1,000 atmospheres. Living creatures exposed to such atmospheres, both inside and out, are perfectly at ease under such conditions. (If, however, a deep-sea animal is lifted to the surface, the internal pressure declines only slightly, while the external pressure declines enormously. Its cells burst, and it dies. We ourselves would die for reverse reasons if pressures upon us were greatly increased.)

If we consider the Earth's interior, the pressures go up still higher, for rock and metal are denser than water, and the depths are greater (the columns of rock and metal weighing down upon layers below are longer than the columns of air weighing down upon the surface of the Earth or the columns of water weighing down upon the sea bottom).

Thus, at a depth of 2,200 kilometers, one third of the way to Earth's center, the pressure is already 1,000,000 atmospheres or a thousand times the pressure in the deepest part of the ocean. At a depth of 4,000 kilometers it is 2,500,000 atmospheres. At the center of the Earth it is possibly as high as 3,700,000 atmospheres. This enormous pressure forces the liquid core to stiffen into solidity at the very center despite its enormous temperature, so that within the central liquid nickel-iron core there is a small central solid nickel-iron core.

Naturally, once again Jupiter displays even more extreme conditions. Its central region holds up columns of material eleven times as deep as Earth's core does (though the material of Jupiter is less dense than our own) and withstands a pressure of as much as 10,000,000 atmospheres.

RESISTANCE TO COMPRESSION

What is it about the material in the interior of the worlds that makes it possible for them to withstand such enormous pressures?

To answer that, let's consider a table on whose surface we have placed an object, say a book. Earth's gravitation serves to pull the book downward. If the book were able to move freely, it would fall in response to Earth's gravitation, and it would continue to fall all the way to the center of the Earth if there were nothing to stop it.

But there is something to stop it: the table. To be sure, the table is also pulled downward, but it is stopped from falling by the floor it rests upon, which in turn is stopped from falling by the walls of the building, which are stopped from falling by the foundations, which are stopped—

If we concentrate on the book and the table only, why does not the book, in response to the Earth's pull, simply fall through the table?

It cannot. The book is made of atoms, and so is the table. The outskirts of all the atoms, both of the book and of the table, are made up of electrons. That means there is an electron surface, so to speak, to the book and an electron surface also to the table.

The two electron surfaces repel each other,

and so much more intense is the electromagnetic force than gravitation that all the pull of the vast Earth cannot force the book through the table against the resistance of those repelling electrons. In other words, the gravitational force is countered by the electromagnetic force, and an equilibrium is achieved in which the book lies quietly on the table, neither passing through it in response to gravitational attraction nor rising above it in response to electromagnetic repulsion.

If the weight of objects on the table is made great enough, however, if enough massive books are piled upon it, the table will break at some weak point; the atoms making it up will pull apart at a point where the electromagnetic cement is weaker than elsewhere.

If the weight is on some other sort of object, a wax block for instance, the molecules of wax under the pressure of the weight will slip and slide over one another very slowly. The wax block will deform, and the weight will sink into the wax —not into the substance, but down past the original surface because the wax will flow outward to get out of the way. (Then possibly it will flow back over the weight.)

Both effects are produced in the Earth under the weight of its own uppermost layers. There are cracks, for instance, that represent weak points in the Earth's crust. In fact, the Earth's crust is made up of a number of large plates forever pulling apart, coming together, and rubbing sideways against one another. A sudden motion of the material on one side of a crack against the material on the other is the equivalent of a sudden break under stress, and earthquakes result. Some distance under the surface, where heat makes the rock more capable of slowly deforming, wax fashion, the heated rock, or *magma*, can squeeze up through weak points in the harder layers above and produce a volcanic eruption.

As one goes deeper and deeper into the Earth's interior, however, there is less chance for cracks and breaks, and deforming becomes slower. Something else must happen to material at great depths and under great pressure. That something else is *compression*.

In the laboratory scientists are most familiar with the effects of increasing pressure in connection with gases. Gases are composed of speeding molecules that are separated from other molecules by distances that are large compared with their own size. If gases are compressed, the molecules are pushed more closely together, and some of the empty space is, so to speak, squeezed out. Gases are easily compressed into smaller volumes by pressure, then. Gases can be compressed to a volume of 1/1,000 their original volume or less before all the empty space is squeezed out and the molecules are in contact.

In liquids and solids, however, the atoms and molecules are in contact already and therefore cannot be compressed as gases are, by having empty space squeezed out. That is why when liquids or solids placed under the kind of pressure that suffices to compress gases, nothing seems to happen to them. Liquids and solids are therefore said to be "incompressible."

This is sufficiently true under ordinary conditions to make it possible for hydraulic presses to work and for steel girders to hold up skyscrapers. Nevertheless, it is not absolutely true.

If pressure is placed on liquids or solids, the

atoms themselves are compressed; the electrons are driven inward toward the nucleus. This is done even under the slightest pressures—that of the book on the table for instance. The outermost electrons are driven inward along the plane of contact; the amount by which the electrons are driven inward under the kind of pressures we encounter in everyday life is so microscopically small, however, as to be immeasurable.

As atoms are compressed and the electrons are driven inward closer to the nucleus, the intensity of the repulsion between the electrons of adjacent atoms (which are also driven together by the pressures) increases. It is rather like compressing a spring that pushes outward more and more forcefully the more it is compressed. In either case a new equilibrium is reached. A pressure from without compresses the atom or the spring until the return push from within increases to the point where it balances the pressure from outside.

Although an immeasurable compression suffices for ordinary pressure, given *enough* pressure the compression of atoms becomes measurable and the electrons are driven inward noticeably. This means that the atoms in substances under pressure take up less room, which means there is more mass in a given volume—which is another way of saying that the density goes up.

We would expect, then, that in Earth's interior the densities of the substances making it up should increase and be higher than they would be if those substances were on the surface under no pressure greater than that of the atmosphere.

Actually the density of Earth's substance does increase with depth and with the pressure upon it.

As soon as Cavendish worked out the mass of the Earth, it was immediately evident that the Earth could not be the same density throughout, that it had to be considerably more dense in the depths than on the surface.

The ocean has a density of 1 g/cm^3, and the rocks of the outer crust, though differing from each other in density, have an average density of about 2.8 g/cm^3. Yet the Earth's overall average density is 5.52 g/cm^3.

Since the outer layers of the Earth are less than 5.52 g/cm^3 in density, the inner layers must be more than 5.52 g/cm^3. To be sure, the interior core of the Earth consists of molten nickel-iron, and that is indeed denser than the outer rock. The density of iron, the major component of the core, is 7.86 g/cm^3 here at the surface. That, however, is not quite enough to account for Earth's average density. What does account for it is the rise in density through the action of pressure and compression.

The Earth's mantle extends from nearly the surface down to a depth of about 2,900 kilometers, about four ninths of the way to the center. Throughout its extent the chemical composition of the mantle doesn't change very much, and a sample of its substance on the surface would have a density of a little over 3 g/cm^3. Its density grows steadily higher with depth, however, and at the bottom of the mantle it is nearly 6 g/cm^3. The average density of the mantle is 4.5 g/cm^3.

At a depth of 2,900 kilometers, one passes from the rocky mantle into the liquid nickel-iron core, and there is a sudden rise in density, since iron is denser than rock. However, although iron has a density of 7.86 g/cm^3 at the surface, under the pressure of the 2,900-kilometer-deep mantle the density of the core at its outer edge is about 9.5 g/cm³. This density rises further as one penetrates deeper into the core, and at the very center of the Earth it is something like 12 g/cm³. The average density of the core is 10.7 g/cm³. Even the maximum density of the core, however, is still only about half the density of osmium at Earth's surface. If the Earth's core were made of osmium, the pressure would bring its density to about 30 g/cm³.

(Earlier in the book I said that if the Earth had an even density throughout, the gravitational pull would decline steadily as we penetrated beneath the surface and would reach zero at the center. Because of the changing density in Earth's interior this is not quite so. So much of the Earth's mass is concentrated in the relatively small liquid core—which contains 31.5 percent of the Earth's mass in 16.2 percent of its volume—that the gravitational pull actually goes up slightly as one penetrates the Earth. In fact, by the time we found ourselves, in imagination, on the boundary between the mantle and the core, the gravitational pull on us would be 1.06 times what it is on the surface. As, however, we penetrated the core, the gravitational pull would finally begin to decrease and would reach zero at the center.)

At the center of the Earth atoms have only about 85 percent of the diameter they would have on the surface. The electrons have been driven in about 15 percent of the way toward the central nucleus, and that small inward push creates enough outward pressure to balance the very worst the Earth's gravitational pull inward can do. This is another indication of how much more intense the electromagnetic force is than the gravitational force.

STARS

We see, then, that all objects up to the size of Jupiter at least are stable, thanks to the electromagnetic force.

To begin with, individual gas molecules, small particles of dust, and larger solid particles that reach the size of pebbles, boulders, and mountains are all held together by the electromagnetic force only. The gravitational force of such small bodies is so small in comparison that it can be ignored.

By the time we begin to deal with objects the size of large asteroids, the gravitational fields set up by these objects are beginning to pull the matter of the objects inward with noticeable force. The inner regions come under measurable gravitational compression, therefore, and this is more and more the case as the objects under consideration grow larger: Moon—Earth—Saturn—Jupiter. In every case the atoms of the object are compressed until the level of compression produces an outward push capable of balancing the inward gravitational pull.

The equilibrium thus established is an essentially permanent one.

Imagine a body like Earth or Jupiter alone in the universe. The gravitational force and the electromagnetic force in such a world would remain at an eternal standoff, and the material structure of the body itself would remain, as far as we know, in its general overall condition forever. There might be minor earthquake quiverings as the substance of the planet made minor adjustments in its position. The planet might slowly cool off till it had no more heat, in the center or on the surface, and its oceans and atmosphere might freeze, but these are what would be called, from an astronomical standpoint, trivial changes.

The equilibrium is not, however, one between equals. Although the electromagnetic force is unimaginably more intense than the gravitational force, it is the electromagnetic force that is the underdog.

The electromagnetic force, huge and intense though it is to begin with, works only through the individual atom. Each individual atom deep in the interior is compressed and can call for no help, so to speak, from its neighbors, who are all equally compressed. When, therefore, the maximum resistance to compression is exerted by one atom, it is exerted by all under the same pressure. If the pressure is further increased, each atom and all the atoms together come to the end of the row.

The gravitational force, however, incredibly weak though it is to begin with, will build up indefinitely as more and more matter is grouped into one place, as each bit of matter adds its own gravitational field to the whole. Though the resistance to compression can reach only a certain limit, the forces producing the compression can increase without limit.

The electromagnetic force resists compression and supports (with groans, we might imagine) the pressures of Earth's layers as they are pulled inward by Earth's gravitational field. They support (with more agonizing groans, in our fantasy) the much larger pressures of Jupiter's more copious layers pulled inward by Jupiter's larger gravitational field.

Well, then, what happens if we pile matter together to make a heap even larger than Jupiter? May there not come a point where as the gravitational field becomes ever more intense and the pressures at the center ever greater, the atoms that must support it all would finally collapse—like a table that breaks at last under a too great weight placed upon it?

But can we honestly say that heaps of matter larger than Jupiter are possible? It may be that, for some reason, Jupiter is as large as an object can grow.

Of course not. Jupiter may be by far the greatest planet we have observed, but we have, near at hand, closer to us than Jupiter is, an object far larger still—the Sun.

The Sun is as much larger than Jupiter as Jupiter is than Earth. The Sun has a diameter of 1,391,400 kilometers, which is 9.74 times as great as Jupiter's diameter. It would take nearly ten Jupiters side by side to stretch across the width of the Sun. Compare this with the eleven Earths side by side it would take to stretch across the width of Jupiter.

And whereas Jupiter is 317.9 times as massive as Earth, the Sun is 1,049 times as massive as Jupiter.

Another indication of the Sun's vast size in comparison with any of the planets, even Jupiter, rests with the matter of surface gravity. At the visible surface of the Sun, the pull of its gravitational field is just 28.0 times that of the Earth's, or 10.6 times that of Jupiter's.

The escape velocity from the surface of the

Sun is 617 km/sec, which is 55 times that of the Earth and 10.2 times that of Jupiter. In fact, even at a distance of 149.5 million kilometers from the Sun's center the escape velocity from the Sun is still 40.6 km/sec.

Since 149.5 million kilometers is the distance of the Earth from the Sun, it follows that the escape velocity from the Sun from a position on Earth is considerably higher than the escape velocity from Earth itself. This means that when a satellite is sent to the Moon, Mars, or Venus at a velocity great enough to free it from Earth's gravitational pull, it is not necessarily freed from the Sun's gravitational pull. Such a satellite may not circle Earth, but it does remain in orbit around the Sun.

So far only two man-made objects have attained velocities that would set them free from the Sun as well as from the Earth, sending them hurtling out of the solar system. These are the two Jupiter probes, Pioneer 10 and Pioneer 11. This was accomplished by skimming the probes around Jupiter and letting its gravitational field accelerate them to the proper velocity (the escape velocity from the Sun being, in any case, smaller at the distance of Jupiter than at our own distance).

There are more important differences between the Sun and Jupiter. Jupiter is much larger than the Earth, but it is still a planet. Both Jupiter and Earth are, at least on the surface, cold, and they would be dark but for reflecting the light of the Sun.

The Sun, however, is a *star*. It shines with a light of its own, bright and blazing.

Is it a coincidence that the Sun is far more

massive than any planet we know and that it is also blazing with light? Or do the two go together?

We might argue that size and light go together and do so in this fashion:

In coming together a world converts the kinetic energy of the infall of its components into heat, as we saw earlier in the book. The larger the world, the greater the internal heat. The Earth is white hot at its center, and Jupiter is far hotter still.

The Sun, then, being much larger than Jupiter, would also be much hotter in the center—hot enough, perhaps, so that the outer region would no longer serve as sufficient insulation to keep the surface cold. We might argue that the internal heat of an object the size of the Sun would be enough to flow outward in sufficient quantity to keep the solar surface at the white-hot temperature of 6,000°C.

The trouble with this view of the Sun and its structure is that it can easily be shown to be an impossible one.

The Sun, after all, is pouring out energy at a vast rate, and it has certainly been doing it for all of recorded history. It seems to have been doing it for very many millions of years into the past, judging from the record of life on Earth through those past ages. Yet if all the energy the Sun had was what it had gained through the kinetic energy of its formation, then it simply would not have had enough energy at its disposal to be the Sun we know.

In 1853 the German physicist Hermann Ludwig Ferdinand von Helmholtz (1821-1894) tried to calculate how much kinetic energy would be

required to supply the Sun's radiation. He decided
that the Sun would have had to contract from a
mass of matter 300 million kilometers across to
its present size in some 25 million years to produce all the energy the Sun has expended in that
time.

At a diameter of 300 million kilometers, however, the Sun would have filled the entire orbit of
the Earth, which could then only be 25 million
years old at the most. But this was impossible.
Geologists and biologists were quite certain that
Earth was much older than that.

This meant that the Sun was actually gaining
energy from some source other than its own contraction, that it was radiating away this energy as
light and heat, and that it could have continued
to radiate for the entire history of the Earth without getting any cooler. Throughout the nineteenth
century, however, no source from which the Sun
might be gaining energy could be worked out
without introducing difficulties that could not be
explained away.

The turning point came at the turn of the century, when the structure of the atom was worked
out. The atomic nucleus was discovered, and it
came to be clear that there is energy packed inside the nucleus in amounts far greater than exists among the electrons, from which the more
common forms of energy are derived.

The Sun is not, therefore, a ball of ordinary
fire at all. It is a ball of nuclear fire, so to speak.
Somewhere in its center the energies made available by the nuclear force, a thousand times more
intense than the electromagnetic force, are somehow being tapped.

DEGENERATE MATTER

The average density of the Sun is 1.41 g/cm³, a value just a trifle higher than that of Jupiter. This is a density associated with liquids or solids made up of the lightest varieties of atoms. It is definitely not associated with gases. Even the densest gas on Earth has a density of only a little over 1/100 that of the Sun.

What's more, the figure of 1.41 g/cm³ represents only the Sun's *average* density. Deep within the Sun its substance, under the huge pressure of the layers above, which are pulled downward by the Sun's enormous gravitation, must be compressed to a density considerably greater than the average.

To be sure, the Sun's outermost layers are clearly gaseous, since for one thing we can see, through the telescope, great gouts of glowing gas shooting upward from the surface. What's more, the surface temperature of the Sun is 6,000°C, and no known substance can remain liquid or solid at that temperature under ordinary pressures.

The interior of the Sun must be considerably hotter than the surface, but the pressures must be enormous. It seemed natural even as late as the 1890s to suppose that under those pressures the solar substance was compressed into white-hot solids or liquids and that this accounted for the Sun's high density. (This is now known to be true of Jupiter.)

Close consideration of the properties of the Sun in the first quarter of the present century, however, made it clear that it behaves as though it were gaseous throughout, even at its very center. This would have seemed absolutely impossible to the scientists of the 1890s, but a generation later it seemed quite natural because by then knowledge had been gained of the interior of the atom. It came to be understood that the tiny atom is a loose structure of particles far tinier still.

This is the way it came to appear:

Atoms are compressed at the center of the Earth, and the expansive force of these compressed atoms is great enough to hold up all the substance of the overlying layers of the planet like so many miniature Atlases. The atoms are even more compressed at the center of Jupiter, and these can therefore hold up the far greater mass of that giant planet.

Even the little Atlases have their breaking point, however. The mass of the Sun, a thousand times as great as that of Jupiter, under the inward pull of an enormous gravitation reach and pass the limits of strength of intact atoms. The pressure at the center of the sun is equal to 100,000,000,000 atmospheres, or 10,000 times that of Jupiter.

The steady accumulation of matter strengthens the gravitational intensity to the point where it overcomes the electromagnetic force that keeps atoms intact, and those atoms, so to speak, cave in.

The electron shells are smashed under pressure, and the electrons can move about unconstrained by the shells. They push together to form a kind of unstructured electronic fluid, taking up far less room than they would as part of shells in intact atoms. As they push together, the electromagnetic repulsion between them increases further; the electronic fluid can withstand a far greater gravitational compression than intact atoms can.

Within the electronic fluid, nuclei can move freely and can approach each other more closely, as closely as chance dictates. They can even collide with each other.

In ordinary atoms, as they exist on Earth or even in the center of Jupiter, the electron shells act as "bumpers." The electron shells of one atom cannot be very far interpenetrated by those of another; and as long as the nuclei must remain at the center of these shells, they are kept relatively far apart. Once the electron shells are smashed and the electrons compress into the more compact electronic fluid, the average separation of the nuclei decreases considerably.

Matter in which the electron shells are broken and in which the nuclei move about in an electronic fluid is called *degenerate matter*. Degenerate matter can be much denser than ordinary matter. It is the nuclei that make up the really massive portion of matter, and it is they that are the true contributors to the mass of any object. If they are forced closer together in degenerate matter than in ordinary matter, there is much more mass per volume in the former and, therefore, a much higher density.

Despite this high density, however, the nuclei, taking up only a millionth of a billionth of the volume of intact atoms, can still move about freely, just as atoms or molecules do in ordinary gases. Degenerate matter, despite its high density, therefore acts as a gas and has properties characteristic of a gas—a "nuclear gas," if you will.

The first discussion of this concept of the Sun as gaseous throughout came in 1907 in a book by the Swiss astronomer Jacob Robert Emden (1862-1940). The idea was fleshed out and given substance in 1916 by the English astronomer Arthur Stanley Eddington (1882-1944).

He reasoned that if the Sun were a ball of gas throughout, with ordinary atoms in the outer layers and smashed atoms in the inner layers, it ought to act like any other gas. When gases are studied in the laboratory, there is always a balance between any force tending to compress the gas and the temperature of that gas tending to expand it.

In the Sun the gravitational pull must therefore also be countered by the internal temperature of the Sun. The size of the Sun's gravitational field and of its compressing effect was known. Eddington set about determining how high the temperatures in the Sun must be to produce an expansive effect that would counter it.

The results were astonishing. The enormous compressions produced by the Sun's gravitation results in a density of the Solar material at the center that must be in the neighborhood of 100 g/cm^3, four times as dense as the densest material on Earth's surface. Yet the Sun, even with so dense a core, behaves as though it were a gas throughout. The central temperature of the Sun is 15,000,000°C. It takes that high a temperature to keep the Sun expanded sufficiently to produce an overall density of only 1.41 g/cm^3 in the face of its gravitation. (The puzzle about that density, you see, is not that it is so great, but that it is so small.)

And what is it that produces so enormous a

temperature at the Sun's core? It was clear by
Rutherford's time that only nuclear energy would
suffice. *Nuclear reactions,* in which nuclei absorb,
give off, and transfer hadrons, produce much
more energy than the *chemical reactions* we are
familiar with, in which atoms absorb, give off,
and transfer electrons. The former involves the
nuclear force, which is much more intense than
the electromagnetic force involved in the latter.

The next question, then, was just which nuclear reactions are involved in powering the Sun.

To answer that question, something had to be
known about the chemical constitution of the
Sun so that one might begin with a reasonable
notion as to which nuclei exist at the center and,
therefore, which nuclear reactions are possible.

Fortunately, the chemical composition of the
Sun can be deduced from an analysis of its light.
Light is composed of tiny waves, and sunlight
consists of a mixture of light of every possible
wavelength.

Different atoms produce lights of particular
wavelengths characteristic only of themselves,
and on occasion they absorb light of exactly those
same wavelengths. Sunlight can be spread out
by an instrument called a *spectroscope* into a
spectrum, in which all the wavelengths are arranged in order.* In the spectrum are thousands
of dark lines representing wavelengths that have
been absorbed by atoms in the Sun's outermost
layers. The positions of those lines in the spectrum can be accurately determined, and from

* We sense different wavelengths of light as difference in color, and
the most spectacular example of a spectrum occurring in nature is the
rainbow.

those positions the various kinds of atoms that did the absorbing can be identified.

As early as 1862 the Swedish physicist Anders Jonas Angstrom (1814-1874) detected the presence of hydrogen in the Sun. Knowledge of the Sun's composition increased steadily, and in 1929 the American astronomer Henry Norris Russell (1877-1957) was able to work out the Sun's composition in considerable detail.

About 90 percent of all the atoms in the Sun, it turned out, are hydrogen, and it therefore seems plausible to suppose that the nuclei in the center must be predominantly hydrogen nuclei, which consist of single protons. Therefore, the nuclear reactions that would be required to supply the vast stores of energy the Sun constantly radiates would most certainly have to involve the hydrogen nuclei. There just isn't enough of any other kind of nucleus to account for all the energy the Sun has been radiating away in its 5 billion years of existence.

In 1938 the German American physicist Hans Albrecht Bethe (1906-) used knowledge gained about nuclear reactions in the laboratory to work out what might be going on in the Sun.

At the great pressures and densities of the Sun's core the hydrogen nuclei—protons—are very close together and are unprotected by intact electron shells. At the enormous temperature of the Sun's core they move with a speed far more rapid than would be possible on Earth. The combination of closeness and speed means that the protons smash into each other very frequently and with enormous force. Occasionally they remain together, *fusing* into a larger nucleus.

The details of what happens may be under

dispute in minor ways, but the overall results seem certain. At the center of the Sun hydrogen nuclei fuse to form helium nuclei, the next most complicated specimen. Four protons combine to form a helium nucleus, made up of four nucleons—two protons and two neutrons.

Here we have, then, a fundamental difference between a planet and the Sun.

In a planet the inward pull of gravitation results in the compression of atoms, which produces a balancing outward push by the electromagnetic force.

In the Sun the much greater inward pull of gravitation can no longer be countered by the atoms' resistance to compression, and the atoms shatter, so to speak, under the pressure. Instead, gravitation is countered by the expansive push of the heat produced by nuclear reactions that are not possible in the lesser temperatures of planetary interiors.

No doubt there is some critical mass below which atom compression is sufficient, and the body is a planet; and above which the central atoms shatter, a nuclear reaction is ignited, and the body is a star. Somewhere in the range of mass between that of Jupiter and the Sun there must be that critical mass.

Undoubted stars are known which are much smaller in mass than the Sun. A star listed in catalogues as Luyten 726-8B is estimated to have only $\frac{1}{25}$ the mass of the Sun, yet we can just see it by the light of its own feeble blaze. Luyten 726-8B is only 40 times as massive as Jupiter, but it is a star and not a planet.

Indeed, Jupiter itself is suspect. It radiates off into space about three times as much energy as it receives from the Sun. Where does that extra energy come from?

It may be that Jupiter is still contracting slightly and that the kinetic energy of that contraction is converted into heat. It may also be that the atoms at the center of Jupiter are at a temperature and pressure that is bringing them to the edge of the breaking point, that a tiny bit of hydrogen fusion is taking place—just enough to account for a little extra heat leakage from the planet.

If that is so, Jupiter is at the edge of nuclear ignition. There's no fear of actual ignition, of course; it is not large enough and will stay forever only at the edge of ignition.

4 WHITE DWARFS

RED GIANTS AND DARK COMPANIONS

THERE IS A DIFFERENCE between planets and stars that in the long run is more crucial than the mere fact that planets are less massive than stars, or that planets are cold and opaque and that stars are hot and glowing.

Planets are in a state of essentially static stability. The equilibrium between gravitation pulling inward and the electromagnetic field of compressed atoms pushing outward is a perpetual standoff. It can, as far as we know, maintain itself forever if there is no outside interference. Were it alone in the universe, Earth might be frozen and lifeless but its physical structure would persist, perhaps forever.

Stars, however, are in a state of dynamic stability, for a star maintains its structure at the cost of something within that is constantly changing. The inward-pulling gravitation is, indeed, essentially changeless, but the outer push of temperature at the Sun's center, which balances that pull, depends on nuclear reactions that consume hydrogen and produce helium. The Sun remains what it is only at the expense of steadily converting 600,000,000,000 kilograms of hydrogen into 595,800,000,000 kilograms of helium *every second.* *

Fortunately there is such an enormous quantity of hydrogen in the Sun that even at this rate of conversion we need not fear anything drastic happening in the near future. The Sun has been consuming hydrogen in its nuclear furnace for some 5 billion years, and even so there is enough left for at least 5 to 8 billion additional years.

But even 5 to 8 billion years is not eternity. What happens when the hydrogen is gone?

As nearly as astronomers can tell now from their studies of nuclear reactions and of the nature of the various stars they can see, it seems that the dwindling of the hydrogen is the prelude to stark changes in a star's structure.

As the Sun, for instance, uses up hydrogen and accumulates helium at the center, the core will contract further as heavier nuclei concentrate the inner portion of the gravitational field still further. The core will become denser and hotter. Eventually the heat of the core will begin to rise rather sharply, and the additional heat will force the outer regions of the Sun to expand enormously.

Although the total heat of the Sun's outer regions will then be considerably greater than it is

---

* The missing 4,200,000,000 kilograms is converted into the radiation that pours steadily out of the Sun in every direction.

now, it will be spread out over a vastly larger surface. Each bit of surface will have less heat than it now does, and the new surface will be cooler than the present surface is. Where the Sun has a surface temperature of 6,000°C right now, the surface of the expanded Sun will be no more than 2,500°C. At that lower temperature it will gleam only red hot. This combination of vast size and ruddy glow gives this stage of a star's life history the name of *red giant*. There are stars that have reached this stage right now, notably Betelgeuse and Antares.

At its fullest extension the red giant into which our Sun will evolve will be large enough to engulf the orbit of Mercury, or even that of Venus.* Earth will then be quite uninhabitable; life on the planet would have become impossible in the early stages of the Sun's expansion. (Perhaps by then mankind, if it still exists, will have left Earth for homes on planets circling other stars, or in artificial colonies built far out in space.)

By the time the Sun reaches its maximum extension as a red giant, it will have been reduced to the last dregs of its hydrogen. The center of the Sun, however, will have by then become hot enough (with a temperature of at least 100,000,000°C) to cause the helium atoms that had been formed from hydrogen atoms over the past eons to fuse into still larger nuclei and those into larger nuclei still until iron nuclei are formed, each made up of 26 protons and 30 neutrons.

---

* Naturally, if a star is larger than the Sun to begin with, it will expand even farther. The star Antares is so large that if it were in the place of the Sun, its giant sphere would include the orbits of Mercury, Venus, Earth, and Mars.

The amount of energy available from the further enlargement of nuclei is only about 6 percent of the amount originally available from the fusion of hydrogen to helium. Once iron is formed, moreover, matters come to a dead end. No more energy from nuclear reactions is available.

After the hydrogen is used up, therefore, and the red giant is at maximum extension, its remaining life as an object powered by nuclear reactions has to be less than a billion years—even considerably less.

And as the nuclear reactions dwindle and fail, there is then nothing to resist the inexorable inward pull of the gravitational field produced by its own mass. Gravitation has been waiting, pulling patiently and tirelessly for many billions of years, and finally resistance to that pull has collapsed, and the bloated Sun, or any red giant, can do nothing but shrink.

Shrink it does, and it is that which puts us squarely on the high road to the black hole, with two stopping points where we must pause en route.

The story of the first stopping point begins with a German astronomer named Friedrich Wilhelm Bessel (1784-1846). He was one of those who tried to measure the distance of stars and was in fact the first to succeed.

Stars have a motion of their own (*proper motion*), but it is very small indeed in appearance because they are so far away. (Think how much more slowly an airplane very high in the air seems to move against the sky as compared with one that is quite low.)

In addition to the proper motion, stars should seem to move in response to the change in angle from which they are seen from the Earth as it moves in its large elliptical orbit around the Sun. As the Earth moves around the Sun in this fashion, a star should mark out, in reflection of this motion, a very tiny ellipse in the sky (provided you subtract the proper motion and other interfering effects). The farther the star, the smaller the ellipse, and if the size of the ellipse (called *parallax*) can be measured by very delicate work at the telescope, the distance of the star can be determined.

In 1838 Bessel announced that he had accomplished the task for a rather dim star called 61 Cygni, which, it turns out, is some 150 trillion kilometers from Earth. Even light, which travels at a speed of 299,792.5 kilometers per second cannot cover such a tremendous distance quickly. It takes light 11 years to travel from 61 Cygni to us; 61 Cygni is therefore said to be 11 *light-years* distant from us.

Bessel went on to try to determine the distance of other stars, and he fixed on Sirius, which for a number of reasons seemed even closer than 61 Cygni. For one thing Sirius is the brightest star in the sky, and this brightness might be due to its relative closeness.

Bessel carefully studied the position of Sirius night after night and noted the manner in which it very slowly moves relative to the other stars in the course of its larger-than-average proper motion. He expected the motion to shift in a certain way that would indicate the formation of an ellipse in response to Earth's motion around the Sun. This ellipse exists, but superimposed on it he detected a wavering that clearly had nothing to do with the manner in which the Earth moves about the Sun.

After a careful analysis of Sirius's puzzling motion Bessel concluded that it moves in an ellipse of its own and completes each turn of that ellipse in about 50 years.

The only thing that can make a star move in an ellipse like that is by having it respond to a gravitational field. Nothing else was known in Bessel's time that could do it, and nothing else is known in our time either. What's more, a gravitational field large enough and intense enough to move a star out of its path and force it into an ellipse large enough to be measured at a great distance must originate in a mass large enough to be another star.

Bessel could not see anything at all in the neighborhood of Sirius that would serve as the source of a gravitational field, yet something had to be there. He decided therefore that there was indeed a starlike mass in the proper place, but it originated from a star that was not blazing but was dark. It was a giant star-sized planet, so to speak. Astronomers therefore spoke of the "dark companion" of Sirius.

Bessel went on to note that Procyon, another bright star, also has a wavy motion, and therefore he concluded it also had to have a dark companion. It even seemed as though dark companions might be fairly common but that this fact was masked by the impossibility of seeing them directly.

Nowadays we would be very suspicious of such a conclusion. We know that any object with a starlike mass *must* ignite into nuclear reactions at the center and blaze if it is to be anything at all like our Sun. To be both starlike in mass and to be dark as well would require a set of conditions that are widely different from those we are familiar with in our Sun.

To Bessel and his contemporaries, however, a dark companion was not mysterious at all. It was a star that, for some reason, had stopped shining. It had used up its entire energy store (whatever that might be, for Bessel had no way of knowing about nuclear reactions) and was rolling on, as large as ever and with as large a gravitational field as ever, but it was cold and dark.

How could Bessel have guessed what an odd object he had discovered? He certainly could not have known its connection with red giants, since they had not yet been dreamed of in his time.

SUPERDENSITY

The darkness of the dark companions ended in 1862, thanks to the work of an American telescope maker, Alvan Graham Clark (1832-1897), Clark was preparing a lens for a telescope ordered by the University of Mississippi just before the Civil War began. (Because of the war it could not be delivered, and it went to the University of Chicago instead.)

When the lens was done, Clark decided to give it a final test by actually using it to look at the sky and see how good a job it would do. He pointed it at the star Sirius in the course of this test and noted a tiny spark of light in its vicinity, something that was not indicated as being there by any of the star maps.

Clark at first assumed that the spark of light

was the result of an imperfection in the lens and
that part of the light of Sirius was somehow
being deflected. Further tests, however, showed
that there was nothing wrong with the lens. Nor
could he do anything that would cause the spark
of light to disappear or change its position. Furthermore, that position happened to be exactly
where Sirius's dark companion was supposed to
be at that time.

The conclusion was that Clark was seeing the
dark companion. It was very dim, only about
1/10,000 as bright as Sirius, but it was not altogether dark. Sirius's dark companion had become
Sirius's dim companion, and it is usually referred
to now as Sirius B, while Sirius itself can be called
Sirius A. Sirius is now called a *binary*, or double
star system.

In 1895 the German American astronomer
John Martin Schaeberle (1853-1924) detected a
spark of light near Procyon. Its "dark companion," was a dim companion, too, and is now called
Procyon B.

Actually this didn't seem to change matters
very much in itself. It meant that if the companions are not totally dead stars, they are at
least dying stars; that although not totally dark,
they are flickering out.

By the time, Schaeberle had seen Procyon's
dim companion, however, affairs were changing.

In 1893 the German physicist Wilhelm Wien
(1864-1928) had shown that the nature of the
light emitted by any hot object (whether a star
or a bonfire) varies with temperature. One can
study the wavelengths of light emitted and the
nature of the dark lines in the spectrum and

come to a firm conclusion as to the temperature of whatever it is that is radiating light.

By Wien's law any star that is flickering out and is therefore cooling down on its way to darkness has to be red in color. Yet Sirius B and Procyon B are white—dim, perhaps, but white.

Just studying the companions by eye wasn't good enough. What was needed was a spectrum so that wavelengths and dark lines could be studied in detail. That was not so easy, since the companions are so dim and are so near the much brighter stars, which tend to drown them out.

Nevertheless, in 1915 the American astronomer Walter Sydney Adams (1876-1956) managed to pass the light of Sirius B through a spectroscope, producing a spectrum he could study. Once he studied that spectrum, there was no doubt that Sirius B is *not* flickering out. It is hot, almost as hot as Sirius A and considerably hotter than our Sun.

Where Sirius A has a surface temperature of 10,000°C, Sirius B has one of 8,000°C. The Sun's surface temperature is only 6,000°C.

From the temperature of Sirius A, we know how bright each little portion of its surface must be —four times as bright as an equal portion of the Sun's surface. We also know how bright the whole surface must be from its apparent brightness when seen from Earth at its distance of 8.8 light years. We can calculate that it must radiate 35 times as much light as the Sun does; and to produce that much light (considering how much each bit of its surface produces), it must be about 1.8 times as wide as the Sun is, or 2,500,000 kilometers in diameter.

(By the turn of the century, you see, astronomers were beginning to realize that the Sun, which had reigned as the most glorious of all heavenly bodies and upon whose energy all living things on earth depend, is, after all, a rather average star and no more. Sirius A is twice as large as the Sun, nearly twice as hot, over thirty times as luminous. But then, we need not feel deprived. If Sirius A were to replace our Sun in the sky, it would be a brilliant light indeed—too brilliant, for Earth's oceans would boil away, and Earth would before long become a dead world.)

The mystery was Sirius B, however. At its surface temperature every bit of Sirius B's surface must be giving off not very much less light than does an equal bit of Sirius A's surface. To explain, then, why Sirius B should be so much dimmer than Sirius A, we must conclude that there is less surface to Sirius B—a great deal less surface. At Sirius B's temperature, it must have a surface only 1/2,800 that of Sirius A.

To have that surface, Sirius B must have a diameter only $\frac{1}{53}$ that of Sirius A, or 47,000 kilometers. If this is so, then Sirius B is only planetary in size, for it is roughly the size of Uranus or Neptune. It has only about $\frac{1}{8}$ the diameter of Jupiter and only $\frac{1}{30}$ Jupiter's volume. It has in fact a diameter only 3.7 times that of the Earth.

Adams's discovery meant that Sirius B was an entirely new class of star—one that is white hot in temperature and also of utterly dwarfish size compared with ordinary stars like our Sun. Sirius B is a *white dwarf* and, it soon turned out, so is Procyon B.

If Sirius B were not only planetary in size but planetary in mass as well, there would be no way in which it could blaze away so hotly. Objects the size and mass of Uranus or Neptune simply do not have the kind of pressures at their centers that would suffice to ignite the nuclear fires.

There was no question, however, of Sirius B having a planetary mass, whatever its size. It could not cause a large star like Sirius A to swerve from its straight-line course, if it were not itself starlike in mass. At least not so marked a swerve.

From the known distance of Sirius A and Sirius B from ourselves, and from their apparent separation in the sky, we can calculate how far apart they are. Sirius A and Sirius B are on the average 3,000,000,000 kilometers apart, so that their average distance from each other is a little larger than that of the planet Uranus from our Sun. However, while Uranus takes 84 years to swing around the sun, Sirius B takes only 50 years to complete its swing around Sirius A.

From this it can be calculated that the intensity of the gravitational fields of Sirius A and Sirius B are 3.4 times those of the Sun and Uranus. This means that Sirius A and Sirius B taken together are about 3.4 times as massive as the Sun and Uranus taken together (or the Sun alone, for Uranus adds so little to the Sun's mass that it can be ignored).

Actually, Sirius B does not swing around Sirius A. The two stars revolve around the center of gravity of the system. You might imagine them as the two ends of a dumbbell whirling about some point, the center of gravity, along the wooden stick connecting them. If the two ends of the dumbbell were exactly equal in mass, the center of gravity would be just midway between them. If one were more massive than the other, the center of gravity would be nearer the more massive one, and in proportion to the amount by which it were more massive.

In the case of the Sun and any of its planets, the Sun is so much more massive that the center of gravity is always sufficiently near the Sun's center to make it reasonably correct to say the planet revolves around the Sun. The same is true when we speak of the Moon revolving around the Earth—since the Earth is 81.3 times as massive as the Moon and the center of gravity of the Earth-Moon system is 81.3 times as close to the Earth, therefore, as to the Moon. The same is true when we speak of any other planet-satellite system among the Sun's family of worlds.

In the case of Sirius A and Sirius B, however, the mass is split more nearly equally, so the center of gravity is well out into the space between them. Both stars circle that center, and both therefore shift their positions considerably as they revolve. (If this weren't so, Bessel wouldn't have noticed a distinct waviness in Sirius's motion across the sky.)

From the orbits of Sirius A and Sirius B the location of the center of gravity of the two stars can be determined. From the position of that center of gravity relative to the two stars it turns out that Sirius A must have 2.5 times the mass of Sirius B. Since the total mass of the two stars is 3.4 times that of the Sun, we see that Sirius A, that gorgeous star in our sky, has by itself 2.4 times the mass of our Sun, while Sirius B, that unnoticeable spark, has a mass just a trifle less than that of our Sun.

That Sirius A should be 2.4 times the mass of our Sun is not surprising at all. After all, it is larger, hotter, and brighter than our Sun. Sirius B, however, is clearly abnormal. With a size of Uranus or Neptune, it has a mass about equal to our Sun's.

That means it must be very dense indeed. Its *average* density must be something like 35,000 g/m^3, which is 3,000 times as dense as the material at the Earth's core, and 350 times as dense as the material at the Sun's core.

At the time that Adams worked out the size of Sirius B this was a real stunner, since it was hard to accept densities of that sort. And yet, four years before Adam's discovery, Rutherford had worked out the structure of the atom and had shown that most of its mass is concentrated in the ultratiny nucleus. Nevertheless, scientists had by no means gotten used to the notion, and the thought of broken atoms, with the parts of it shrinking far closer together than is ever possible in intact atoms, was hard to swallow. There was considerable skepticism, therefore, about the possibility of the existence of such white dwarfs.

EINSTEIN'S RED SHIFT

Soon after Adams's discovery, however, a possible way of checking the matter from a completely different direction was worked out.

In 1915 the German Swiss physicist Albert Einstein (1879-1955) published his general theory of relativity. This represented an entirely new outlook on the universe as a whole. According to this new theory there ought to be some phenomena that could be observed that would not be possible if the older outlooks were correct. For instance, when light is radiated by a massive body, the strong gravitational field of the body should, according to general relativity, have some effect on the light.

Einstein, building on the work done in 1900 by another German scientist, Max Karl Ernst Ludwig Planck (1858-1947), had shown that light could be viewed as consisting not only of waves, but of waves that are gathered into packets that in some ways act as particles. These light-particles are called *photons*, from a Greek word meaning "light."

Photons have a mass of zero when at rest and therefore do not act as a source of a gravitational field, nor do they respond to one in the ordinary fashion. However, photons are never at rest but always travel (in a vacuum) at a particular precise speed—299,792.5 kilometers per second. (So do all other massless particles.) When traveling at this speed, photons possess certain energies; and although the action of a gravitational field cannot alter the speed of photons in a vacuum (nothing can), it can shift the direction in which the light is traveling, and it can decrease the energy.

The shifting of direction was observed in 1919. On May 29 of that year a total eclipse of the Sun was visible from Principe Island, off the coast of Africa. Bright stars were visible in the sky near the eclipsed Sun, and their light on its way to Earth skimmed past the Sun. Einstein's theory predicted that this light would be bent very slightly toward the Sun as it passed, so that the stars themselves, sighted along the new direction, would seem to be located very slightly farther from the Sun's disk than they really were. The positions of the stars were carefully measured during the eclipse and then again half a year later, when the Sun was in the opposite half of the sky and could exert no effect at all on the light from those same stars. It turned out that the light behaved as Einstein's theory had predicted it would, and that went a great way toward establishing the validity of general relativity.

Naturally astronomers were anxious to make further checks on the theory. What about the loss of energy of light in a gravitational field? Light leaving the Sun must do so against the pull of solar gravitation. If the photons were ordinary particles with mass, their velocities would decrease as they rose. Since photons have a rest mass of zero, that doesn't happen, but each photon loses a little of its energy just the same.

This loss of energy ought to be detected in the Sun's spectrum. The longer a wavelength a particular photon has, the smaller its energy. In the spectrum, where light is arranged in order of wavelength from violet (with the shortest wavelength) to red (with the longest), there is a smooth progression from the high energy of violet to the low energy of red.

If sunlight loses energy because it rises against the pull of gravitation, every bit of it should end up slightly closer to the red end of the spectrum than it would if there were no gravitational effect. Such a *red shift* could be detected by studying the dark lines in the solar spectrum and comparing their positions with the dark lines in the spectra of objects that are subjected only to small gravitational effects—in the spectra of glowing objects in laboratories on Earth for instance.

Unfortunately there was no point in looking

for this Einstein red shift in the solar spectrum because the effect is so tiny that even the Sun's mighty gravitational field won't produce enough to measure.

But then Eddington (who was working out the internal structure of the Sun and who was very enthusiastic about the theory of relativity) pointed out that if Sirius B is really both as massive and as tiny as it seemed to be, that might be the answer. It is not just the overall gravitational pull that affects light as much as it is the intensity at the surface, where the light is given off and where it makes its initial jump into space.

Now, the intensity of the Sun's gravitational field is 333,500 times that of the Earth's, but the Sun's surface is so far from its center that the Sun's surface gravity is only 28 times that of the Earth.

What about Sirius B? It has the mass of the Sun compressed down into an object the size of Uranus. It has the same gravitational intensity as the Sun has, but you can get much closer to the center of Sirius B by standing on its surface (in imagination only, of course) than you could ever get to the Sun's.

The surface gravity of Sirius B is, therefore, about 840 times that of the Sun and 23,500 times that of the Earth. The Einstein red shift should be much more pronounced in the light leaving Sirius B than in the light leaving the Sun.

Eddington suggested to Adams, who was the Sirius B expert, that he study the spectrum of its light again to see if he could detect the red shift. In 1925 Adams tried the experiment and found that indeed he *could* detect the red shift and precisely to the extent that Einstein's theory had predicted.

Not only did this provide another important verification of general relativity, but if the theory was correct, it provided a strong piece of evidence that Sirius B is indeed as massive and as tiny as Adams had maintained, since only so can it have enough surface gravity to produce the red shift that was observed.

In 1925, therefore, the existence of white dwarfs had to be accepted. There has been no doubt about it since.

The enormous surface gravity of Sirius B implies an enormous escape velocity. From the surface of Earth a missile hurled into the sky with no source of energy other than the initial impetus must start with a minimum velocity of 11.23 km/sec if it is to leave Earth permanently. From the surface of the Sun the escape velocity is 617 km/sec. From the surface of Sirius B the escape velocity is about 3,300 km/sec.

Even 11.23 km/sec is a rapid velocity by earthly standards. A velocity of 3,300 km/sec is, however, enormous. It is $\frac{1}{90}$ as fast as the speed of light.

FORMATION OF WHITE DWARFS

Let's look again, now, at what will happen after our Sun reaches the red giant stage and uses up all the nuclear energy in its interior. The gravitational pull, being then unopposed by the expansive effect of heat, will begin to shrink the Sun (as it seems to do now with other stars that are in that stage) to some point where gravitation is opposed by something other than heat.

As it shrinks, it will gain density till it reaches the point where it might be composed of intact atoms in contact, as planetary bodies such as Earth and Jupiter are. A star-sized mass, however, produces a strong enough gravitational field to smash such intact atoms. Thus, the shrinking will continue. If it is to be stopped at all, that stopping must be done by the subatomic particles that make up atoms.

What are those subatomic particles, and in what manner do they change as the Sun (or any other star) ages?

To begin with, the Sun, or any star, is mostly hydrogen. Hydrogen consists of a nucleus made up of a single positively charged proton that is balanced by a single negatively charged electron making up the rest of the atom.

As the Sun ages, its hydrogen little by little undergoes fusion, four hydrogen nuclei fusing to form a single helium nucleus. Since a helium nucleus is made up of two protons and two (electrically uncharged) neutrons, we may say that when all the hydrogen has fused and is gone, half the protons in the star have changed to neutrons. As helium nuclei undergo further fusion during red-giant formation until finally iron nuclei are formed, a few more protons are converted to neutrons, and in the end the star is a $^{45}\!/_{55}$ mixture of protons and neutrons.

What happens to the electrons meanwhile?

Every time a positively charged proton is converted to an uncharged neutron, something has to be done with that positive charge. It cannot vanish into nothing all by itself. Instead, it is

ejected from the fusing nuclei along with a minimum amount of mass. This minimum amount of
mass is enough to produce a particle exactly like
the electron except that it carries a positive electric charge instead of a negative one. This positively charged electron is called a *positron*. For
every four protons fused into a helium nucleus
two positrons are formed.

Once a positron is formed, it is sure to collide
with one of the electrons present in the Sun (and
in all ordinary matter) in overwhelming numbers. Although a positive electric charge cannot
disappear all by itself and a negative electric
charge cannot so disappear either, the two may
cancel each other if they meet. When a positron
and an electron collide, there is a *mutual annihilation* of both electric charge and mass, and the two
are converted into energetic photos called *gamma
rays* which possess neither electric charge nor
mass.

In this way about half the electrons in the Sun
will have been destroyed in the course of its lifetime as a normal star. The half that remains will
be enough to balance the half of the protons that
have remained as such.

In the conversion from protons to neutrons and
in the mutual annihilation of electrons and positrons enough mass is lost to be converted into all
the vast quantities of radiation the Sun emits in
its lifetime as a hydrogen-fusion reactor. Additional mass is lost because the Sun is always
giving off a stream of protons in all directions,
the so-called *solar wind*.

All this loss is trivial compared to the Sun's
total mass. By the time the Sun, or any star that
exists in isolation, has completed its red-giant

period and is ready to shrink, it can have retained as much as 98 percent of its original mass; it is this mass that now begins to shrink.

Electrons, protons, and neutrons all have wave properties as well as particle properties. The greater the mass of a particle, the shorter the waves associated with it and the more pronounced the particle properties. The less the mass, the longer the waves and the more pronounced the wave properties.

Protons are much more massive than electrons —1,836 times more massive. Neutrons are 1,838 times more massive than electrons. Protons and neutrons are associated with very tiny waves and are pronounced particles of extremely small size. The electron is associated with comparatively long waves and it therefore takes up much more space than protons or neutrons.

As a star collapses, then, past the limit marked by intact atoms, it is the comparatively bulky electrons that are brought together in contact, so to speak, first.

Electrons driven into contact are much more compactly packed than they would be in intact atoms. Thus, Sirius B and the Sun have about equal masses, but Sirius B takes up only 1/27,000 the space that the Sun takes up. (It's something like the difference in the space taken up by a hundred intact Ping-Pong balls and the space taken up by those same Ping-Pong balls broken up into flakes of plastic.)

Nevertheless, even after the electrons have been brought into contact, the much smaller (but more massive) protons and neutrons, and atomic nuclei made up of them, still find plenty of room for movement. These nuclei are much closer together than they would be if they were part of intact atoms, but they are still far enough apart so that the distances between them are very large in comparison with their own size.

As far as the nuclei are concerned, a white-dwarf star, dense as it is, is still mostly empty space. In Sirius B, for instance, which might almost be considered as a continuous electronic fluid, the nuclei take up only 1/4,000,000,000 its volume. The nuclei, therefore, show the properties of gases.

Naturally a white-dwarf star is not of even structure all the way through, any more than any other massive object is. There is increasing pressure as one moves, in imagination, from the surface to the center.

A white dwarf has an almost normal skin, an outermost layer of intact atoms that are pulled down hard by the intense gravitational pull at the surface but don't have the weight of other layers above them. A number of different kinds of atoms can exist in this white-dwarf "atmosphere"— even a small amount of hydrogen that has somehow, through all the life of the star, yet escaped fusion because those particular atoms never happened to make up part of the stellar depths. The atmosphere may be only a couple of hundred meters thick.

As we imagine ourselves sinking down into the material of the white dwarf, these atmospheric atoms gradually break down into electrons and nuclei, both moving freely. There, small dregs of nuclear reactions go on till all the hydrogen is used up. As we continue to move downward, the electrons move into contact and begin to resist further compression. The more tightly they are compressed, the more strongly they resist further compression, and it is this resistance that finally brings a halt to the contraction of the star at the white-dwarf stage.

At the core the material of the white dwarf is considerably denser than average for the whole star. The central density may be as high as 100,000,000 g/cm³.

When a white dwarf is formed, it is very hot indeed, since the kinetic energy of the infall has been turned into heat. A white dwarf freshly formed can have a surface temperature in excess of 100,000°C.

As the white dwarf radiates heat into surrounding space, however, its energy content must decrease, and very little of that decrease can be made up for by the nuclear reactions in the dregs of fairly normal matter that at first remain in its outer layers. Gradually the white dwarf cools down. There are old white dwarfs known with surface temperatures of not more than 5,000°C.

This loss of heat does not seriously affect the structure of the white dwarf. Ordinary stars would collapse if they lost heat, since it is the heat produced in the center that keeps them expanded against the contracting pull of gravity. A white dwarf resists the gravitational inpull by the outward push of the compressed electrons, and this does not depend upon heat. The electrons resist further compression as efficiently when cold as when hot.

Presumably, then, the loss of temperature will continue, with no significant change in white-dwarf structure, until the white dwarf is no longer hot enough to glow. It becomes a *black dwarf* and will continue to cool further over the eons until its energy content is only the average for the entire universe—a few degrees above absolute zero.

This is a very slow process, and the entire lifetime of the universe to date has not been long enough to have witnessed the total drainage of energy from any white dwarfs. All the white dwarfs that have ever formed are still shining today, but given enough time, they will fade.

So far, then, in this book we have discussed two kinds of eternal objects—objects, that is, that can resist the inward pull of gravity for indefinitely long periods of time. There are planetary objects, which are small enough in mass never to have started a nuclear fire, and where gravitational compression is forever balanced by the outward push of compressed intact atoms at the center.

There are also (or will be someday) black dwarfs, which have sufficient mass to have started a nuclear fire but which, in time, have burned out, and where gravitational compression is forever balanced by the outward push of compressed electrons.

All the objects we see in the sky outside our own solar system, plus the Sun within our solar system, are *not* eternal objects. The ordinary stars we see are temporary structures burning their way down to black-dwarf status (or, as we shall see, other, even stranger objects) at last.

We can also see clouds of dust and gas in interstellar space, but under the pull of their own gravitational field much of those clouds will eventually condense to form stars and work their way down to the black-dwarf status, also. Some of the clouds may condense into bodies too small in mass to ignite the nuclear fire, and they will then be *planetary bodies*. If any of the cloud escapes condensation and joins the thin vapor of individual atoms, molecules, and dust particles that spread out between the stars and galaxies, then these may be considered separate ultratiny planetary bodies.

We remain, then, with planetary bodies and with black dwarfs as the two classes of eternal objects in the universe that we have so far discussed in this book.

Several hundred white dwarfs have been observed, and that doesn't seem like much among the billions upon billions of stars in the sky. Remember, however, that white dwarfs, although bright for their size, are nevertheless dim on the whole. They are only 1/1,000 to 1/10,000 as luminous as are average ordinary stars and therefore cannot be seen unless they are very close to us.

We see so few white dwarfs because at usual stellar distances, where ordinary stars are still bright enough to see and study, white dwarfs are too dim to recognize, or perhaps even to see. The only way we can really judge the number of white dwarfs, then, is by studying the immediate neighborhood of the Sun.

In the space within 35 light-years of the Sun, for instance, there are about 300 stars. Of these, 8 are white dwarfs. If we assume that this is about the usual proportion in space generally (and we have no reason to think it isn't), then we can say that somewhere between 2 and 3 percent of all stars are white dwarfs. There may be as many as 4 billion white dwarfs in our galaxy alone.

5 EXPLODING MATTER

THE BIG BANG

WHY ARE THERE as many white dwarfs as there are? Why are there 4 billion in our galaxy alone?

After all, a star doesn't become a white dwarf till it has used up all its nuclear fuel, and our Sun, for instance, still has enough nuclear fuel to last it for billions of years. This may also be true of countless numbers of the 135 billion stars making up our galaxy. Why, then, have 4 billion of those stars run out of fuel, expanded, and then collapsed?

Or suppose we look at it from the other side. Why are there so few white dwarfs? If billions of stars have used up their nuclear fuel and collapsed, why have not all the stars done so?

In order to answer these questions we must know first of all how old the universe is and, therefore, how long ago stars were formed. We might then get an idea as to how long they have been fusing nuclei and how much remains to be fused.

But how can we possibly tell the age of the universe?

The answer to that came, quite unexpectedly, out of a consideration of the spectra of stars.

By studying the spectra of stars it is possible to tell whether a particular star is moving away from us or toward us, and, in either case, how quickly. If the spectral lines are shifted toward the red end of the spectrum, the star is moving away from us. If the spectral lines are shifted toward the violet end of the spectrum, the star is moving toward us.

Of course, we might ask how one can tell whether the red shift of the lines is caused by motion away from us or by a gravitational effect as described in the previous chapter. The answer to that is that most stars aren't dense enough to produce a measurable red shift resulting from a gravitational effect. Therefore, unless there is reason to believe the contrary, every observed red shift is taken to be due to motion away from us.

Naturally, some stars move away from us, and some move toward us, so that red shifts and violet shifts are about equal in number.

Beginning about 1912, however, astronomers began to study the spectra of the galaxies (which are vast and distant collections of millions, or billions, or even trillions of stars similar to our own Milky Way Galaxy) that lie beyond our own. By 1917 it was clear that all but a couple of the closest galaxies show a red shift and are therefore receding from us. What's more, those red shifts are larger than the ones associated with the stars of our own galaxy.

As more and more galaxies were studied, it turned out that *all* the galaxies (except that same couple of the closest) have a red shift and that the size of this red shift increases steadily the farther the galaxies are from us.

Taking all this into account, the American astronomer Edwin Powell Hubble (1889-1953) in 1929 advanced what is called Hubble's law. By this rule the rate at which a galaxy is receding from us is directly related to its distance from us. That is, if galaxy A is receding from us at 5.6 times the velocity of galaxy B, then galaxy A is 5.6 times as far from us as galaxy B is.

The rate of increase of the galaxies' speed of recession with distance isn't easy to determine. At first astronomers thought that the speed increased rather quickly, but newer data has made it seem that the increase is much smaller than had at first been thought. At present astronomers estimate that the speed of recession goes up 16 kilometers per second for every million light-years of distance. For example, a galaxy that is 10,000,000 light-years away from us is receding at a velocity of 160 km/sec, one that is 20,000,000 light-years away from us is receding at a velocity of 320 km/sec, one that is 50,000,000 light-years away from us is receding at a velocity of 800 km/sec, and so on.

But why should this be? Why should all the galaxies be receding from us, and why should this speed of recession be proportional to distance from us? What makes us the key to the behavior of the universe?

We aren't!

As early as 1917 the Dutch astronomer Willem de Sitter (1872-1934) showed that from a theoretical standpoint, using the equations of general relativity, the universe ought to be expanding. To be sure, individual galaxies and sometimes a cluster of anywhere from dozens to thousands of galaxies are held together by gravitational pull. But galactic units (either single galaxies or clusters of them) that are separated from their neighbors by so great a distance that gravitation is too weak to influence them sufficiently, partake of the general expansion of the universe. This means that individual galactic units are all moving apart from one another at some constant velocity.

From a viewpoint on *any* one galaxy it would seem that all other galaxies (except those that are part of the home cluster, if there is one) are receding. What's more, the constant velocity of expansion builds up with distance, so we would end with Hubble's law no matter which galaxy we lived in.

If the galactic units spread out farther and farther from one another as time moves forward and the universe grows older, then, if we look backward in time (like turning a movie film so that it runs the other way) we would see the galactic units coming closer and closer to each other. The universe is more compact, in other words, the younger it is; and if we go far enough back in time, we can see how all the galaxies must have been crushed together in one vast collection of matter.

In 1927 the Belgian astronomer Georges Lemaître (1894-1966) suggested that this was actually so—that a certain number of billions of years ago the matter of the universe was all in one place and formed a structure he called the *primeval atom*. Others have called it the *cosmic egg*.

How long the cosmic egg existed, or how it was formed, Lemaître did not venture to guess, but at some moment it must have exploded. This must surely have been the greatest explosion the universe had ever experienced; it was the explosion that created the universe as we know it. The Russian-American physicist George Gamow (1904-1968) called it the *big bang*.

Out of the vast outward-speeding fragments of the cosmic egg, the stars and galaxies formed eventually, and it is because of the still felt outward force of the big bang that the universe is expanding even today. In the last half century evidence for the big bang has accumulated, and nowadays almost all astronomers accept the thought that that is how the universe started.

The big question, though, is when the big bang took place. Astronomers know (or think they know) just how rapidly the universe is expanding now. If they assume that this rate of expansion has always been the same and will always remain the same, then, if we look forward in time, the universe will just expand forever and ever; the galactic units will separate farther and farther. Finally an astronomer looking out at the universe from Earth will see only our own galaxy and those other galaxies that form part of our Local Cluster. Everything else will be too far off to see.

On the other hand, if we look backward in time and assume that the universe will contract steadily at a uniform rate, it will come together into the primeval atom 20 billion years ago.

However, the various galaxies do exert a gravi-

tational force on one another. This may not be
sufficient to prevent the expansion, but it will
tend to slow it. That means that as we look into
the future, the rate of expansion will grow
smaller and smaller, and it will take longer than
we think before all the distant galaxies outside
the Local Cluster are lost to sight. Similarly, it
means that as we look back into the past, the
galaxies come together faster and faster as gravitational pull becomes more and more important.
Therefore, the time of the cosmic egg and the big
bang must be less than 20 billion years ago.

We are not sure by exactly how much the
gravitational force in the universe is slowing the
rate of expansion. It depends on how much matter there is (on the average) per volume of space
—the average density of matter in the universe
in other words.

If the density is high enough, then the slowing
effect is great enough to bring that rate of expansion to zero eventually. The expanding universe will eventually be brought to a halt. Once
that happens, the universe under the pull of its
own gravitational forces will begin to contract—
at first very slowly, then faster and faster, till the
cosmic egg is formed and explodes again. This
can happen over and over again, and we will
have an *oscillating universe*. The American astronomer Allan Rex Sandage (1928-) has suggested that a cosmic egg forms and explodes
every 80 billion years.

If the density of matter in the universe is just
barely high enough to bring the galaxies to a halt
(a density equal to 6×10^{-30} g/cm^3, or about one
proton or neutron in every 350,000 cubic centimeters of space), then the expansion is slowing

at such a rate that the big bang must have taken place about 13.3 billion years ago.

Actually astronomers are not yet certain just how dense the matter of the universe is, on the average, so that we can't be sure exactly when the big bang took place or whether the universe is oscillating or not. At the present time the general feeling is that the average density is not high enough for oscillation, so that the big bang must have taken place some time between 13.3 and 20 billion years ago.

In this book let's make the reasonable assumption (subject to change as additional evidence is gathered) that the universe is 15 billion years old.

If the universe is 15 billion years old, that means the stars themselves can't be more than 15 billion years old.

They might be younger than that, however. The Sun, for instance, must be younger than that, or by now it would have consumed its nuclear fuel, expanded to a red giant, and collapsed to a white dwarf.

Can it be, then, that white dwarfs are the remnants of very ancient stars that have been shining since the beginning of the universe, while those stars that still shine by nuclear fusion were formed much later and are much younger?

There is possibly something to that, but it can't be the whole answer. *Many* stars must have been formed after the big bang, and if they had all reached the white-dwarf stage by now, there would be far more white dwarfs in our galaxy than in point of fact there are. Then, too, consider Sirius A and Sirius B. It seems logical to suppose that the two stars of a binary were

formed at the same time (just as the Sun and
the planets must have been formed at essentially
the same time), and yet one is a white dwarf
and one isn't.

Can it be that age is not the only factor that
counts? Do some stars burn nuclear fuel more
slowly than others? Or do some stars have more
nuclear fuel to begin with than others? In either
case, do some stars take longer to reach the stage
of collapse than others?

The answer to this question, too, came from
the studies of spectra.

THE MAIN SEQUENCE

To begin with, a star is born out of a mass of
dust and gas, which swirls slowly and which under its own gravitational pull comes slowly together. As this mass of dust and gas (spread
through space in the wake of the big bang) comes
together, the gravitational pull becomes more and
more intense, so the process hastens.

As the cloud condenses, the temperature and
pressure at the center get higher and higher
until finally they become high enough to break
down the atoms at the center and initiate nuclear fusion. At this moment of nuclear ignition
the developing star is born.

The period of condensation is not very long
compared to the total multibillion-year lifetime of
a star. The larger and more massive the cloud is
to begin with, the stronger the gravitational pull
at all stages and the less the time of condensation. A star the mass of our Sun might take
thirty million years to reach nuclear ignition,

while a star with ten times the mass of the Sun might condense to nuclear ignition in only ten thousand years. On the other hand, a star with only one-tenth the mass of the Sun might take a hundred million years to ignite.

Naturally, the stars we see in the sky have already reached nuclear ignition. Once they have reached it, they continue to produce and radiate energy at very much the same rate for a long period. The actual rate at which any star produces and radiates energy depends on how massive it is.

When Eddington worked out the temperatures inside a star, he realized that the more massive a star, the stronger the gravitational force pulling it together. This meant that the more massive a star, the greater the internal temperature required to force it to remain expanded in the face of gravity. The greater the internal temperature, the more energy will be produced and the more will radiate away from the star. In other words, the more massive a star, the more luminous it will be. Eddington's rule is called the *mass-luminosity law*.

If we study the stars we see, we find that they form a regular sequence from very massive, very luminous, very hot stars through stages of smaller and smaller mass, luminosity, and heat down to stars of very small mass, very little luminosity, and quite cool surfaces. This sequence is called the *main sequence*, since it constitutes about 90 percent of all the stars we know of. (The other 10 percent are unusual stars, such as red giants and white dwarfs.)

The spectra of the stars of the main sequence form a sequence of their own. As one moves along the main sequence toward cooler and cooler stars, the spectra reflect the steadily lower temperatures in the nature of the dark lines they contain. The stars can therefore be divided into *spectral classes* according to the dark-line pattern.

The spectral classes into which the main-sequence stars are divided are O, B, A, F, G, K, and M. Of these O is the most massive, the most luminous, and the hottest; while M is the least massive, the least luminous, and the coolest. Each spectral class is divided into subclasses numbered from 0 to 9. Thus we can speak of B0, B1, B2, and so on until we reach B9, which is followed by A0. Our own Sun is of spectral class G2.

In Table 9 the mass and luminosity of stars are listed by their spectral class.

Are these stars all equally common?

The answer is no.

TABLE 9—The Main Sequence

Spectral Class	Mass (Sun = 1)	Luminosity (Sun = 1)
O5	32	6,000,000
B0	16	6,000
B5	6	600
A0	3	60
A5	2	20
F0	1.75	6
F5	1.25	3
G0	1.06	1.3
G5	0.92	0.8
K0	0.80	0.4
K5	0.69	0.1
M0	0.48	0.02
M5	0.20	0.001

In the universe generally, large objects are always exceptional and are less common than small objects of the same category. There are fewer large animals than small animals (compare the number of elephants to the number of flies), fewer large rocks than small grains of sand, fewer large planets than small asteroids, and so on.

We might expect, then, that there are fewer large, massive, and luminous stars than there are small, light, and dim stars, and we would be right. The surveys that astronomers have made of the stars they can see and the deductions they have made from these surveys lead them to suppose that nearly three fourths of all the stars in our galaxy fall into spectral class M, the dimmest of all. The results, in detail, are presented in Table 10.

TABLE 10—Spectral-Class Frequency

Spectral Class	Percentage of Stars	Number of Stars in Galaxy
O	0.00002	20,000
B	0.1	100,000,000
A	1	1,200,000,000
F	3	3,700,000,000
G	9	11,000,000,000
K	14	17,000,000,000
M	73	89,000,000,000

(We can assume, of course, that whatever holds true of our galaxy holds true for the vast majority of other galaxies as well. We have no reason to think that our own galaxy is particularly unusual.)

The next question is whether the stars of the various spectral classes take different times to consume their nuclear fuel, and whether some therefore remain on the main sequence longer than others and delay the inevitable expansion and collapse.

If we assume, for instance, that all stars begin their careers with a constitution that is mostly hydrogen, the chief nuclear fuel, then we can see that the more massive a star, the larger the supply of fuel it has. An O5 star, with 32 times the mass, and therefore the supply of nuclear fuel, that the Sun has, *might* (we could assume) take 32 times as long to consume its fuel and would therefore remain quietly on the main sequence 32 times as long as our Sun would, and, for that matter, 160 times as long as an M5 star would.

However, stars don't consume nuclear fuel at the same rate regardless of their masses. The more massive a star, the more powerfully its own gravitational field compresses its matter and the hotter its core must be to balance that gravitational compression. The hotter the core must be, the more fuel must be consumed per second to maintain that temperature. In short, the more massive a star, the more rapidly it must consume its nuclear fuel.

Eddington was able to show, in fact, that as we progress from less massive to more massive stars, the rate at which they must consume their nuclear fuel increases much faster than the nuclear fuel supply does. In short, though an O5 star may possess 32 times as much nuclear fuel as the Sun, that O5 star must consume nuclear fuel over 10,000 times as quickly as the Sun, and must therefore consume its greater supply of nuclear fuel much sooner than the Sun consumes its smaller supply. Reasoning in the same way, the

Sun must use up its nuclear fuel much more quickly than a dim M5 star with only one fifth the supply that the Sun has.

In short, the more massive a star, the shorter it stays on the main sequence and the sooner it becomes a red giant and then collapses. The lifespan of the various spectral classes are given in Table 11.

TABLE 11—Lifespan on the Main Sequence

Spectral Class	Lifespan (years)
O	1,000,000 or less
B0	10,000,000
B5	100,000,000
A0	500,000,000
A5	1,000,000,000
F0	2,000,000,000
F5	4,000,000,000
G0	10,000,000,000
G5	15,000,000,000
K0	20,000,000,000
K5	30,000,000,000
M0	75,000,000,000
M5	200,000,000,000

Since it is the larger, and less common, stars that collapse first, here is one explanation for the relative rarity of white dwarfs. No star of spectral class K or M, which together make up 87 percent of all stars, has had a chance to use up its nuclear fuel yet, even if each had been burning and radiating ever since the big bang. Only the O, B, A, F, and some of the G stars can possibly have left the main sequence, and they make up less than 10 percent of all stars.

Even so, we have not entirely explained the rarity of white dwarfs. If all the stars in the Galaxy had been formed soon after the big bang and none had been formed since, there would be no stars in the Galaxy larger and more luminous than the smaller G-class stars. The brighter ones would all have expanded and collapsed. But this is not so. There are extraordinarily bright stars in the sky right now—even O-class stars.

Clearly, the bright stars that now exist cannot have existed throughout the entire lifetime of the universe so far. They must have been formed comparatively recently. Our own Sun (spectral class G2) mut be far younger than the universe, or it would be a white dwarf right now. As a matter of fact, it seems to have been formed about 5 billion years ago, when the universe was already 10 billion years old. And there are places in the Galaxy where stars seem to be contracting toward nuclear ignition right now; and there will be stars forming a billion years from now.

For a long, long time there will remain luminous, shortlived stars in the sky, coming and going, while the dwarf stars shine on steadily.

Still, if we assume the universe will expand forever, then eventually all stars, even the smallest, will consume their nuclear fuel, expand, and collapse. And many trillions of years from now we might suppose that the universe would consist of only two types of dark "eternal" bodies—black dwarfs that are the cinders of stars and black planetary objects that were never stars.

But if we assume this is the end, will we be right? Does every object large enough to become a star end as a white dwarf cooling into a black dwarf? Or are there objects in the universe even stranger than the white dwarf?

Yes, there are odder objects still on the horizon. Remember we are heading toward the black holes.

PLANETARY NEBULAS

When a star collapses to a white dwarf, its mass, under the influence of its own gravity, pulls together and contracts, growing smaller and smaller until the compressed electronic fluid at the core becomes resistant enough to further collapse to bear up the weight of the layers of matter above it.

The more massive a collapsing star, the more forcibly it will shrink and the more tightly it will compress the electronic fluid.

To make still another analogy, this is rather like the situation with the tires that hold up an automobile. The weight of the automobile compresses the air within the tires. The outward push of the air in the tires becomes greater the more it is compressed, so that eventually it is compressed enough to bear the weight of the automobile. If you then load baggage into the automobile, the air in the tires is compressed further until there is enough outward push to hold up the additional weight. The more weight there is, the tighter the air within the tire is compressed.

If we think of this in connection with a star, we can see that it is quite likely that the more massive a white dwarf is, the smaller it must be in size. Thus, a white dwarf star called Van Maanen 2 is only three-fourths as massive as Sirius B. Therefore, it does not compress as tightly and has a diameter about equal to that of Jupiter, or three times that of Sirius B. On the other hand, some comparatively massive white dwarfs are no larger in volume than our Moon.

But how massive, and how small, can a white dwarf get? After all, if we load more and more weight on a car, the time will come when the material of the tires will not be strong enough to withstand the greater and greater compression of the air. Eventually there comes a point where the tire blows.

Is there also a point where the core of the white dwarf simply cannot hold up the mass pressing down upon it.

The question was taken up by an Indian-American astronomer, Subrahmanyan Chandrasekhar (1910-). In 1931 he was able to show that there is a certain critical mass (*Chandrasekhar's limit*) beyond which a white dwarf cannot exist, since the electronic fluid at that point cannot support the weight no matter how compressed it is. The core of such a star will simply collapse inward.

The critical mass, Chandrasekhar showed, is 1.4 times that of the Sun. The limit could be a little higher if a white dwarf were rotating rapidly, since then centrifugal force would help lift some of the mass upward. White dwarfs, however, do not seem to rotate fast enough to make this a significant factor.

Chandrasekhar's limit is not very high. All the stars of spectral classes O, B, and A, together with the more massive stars of spectral class F, have masses that are more than 1.4 times that of the Sun. These are also the stars with the shortest lifetimes, and examples of such stars formed in the early days of the universe have surely by now expanded and collapsed. Into what did they collapse? Could some of them have collapsed into very massive white dwarfs well beyond Chandrasekhar's limit—thus proving Chandrasekhar's analysis to have been in error?

Conceivably so, but the fact is that all the white dwarfs studied have proved to have masses less than Chandrasekhar's limit, and the more such we find, the better the limit looks.

Another alternative is that stars more massive than Chandrasekhar's limit might at some stage before or during their collapse have lost some of their mass.

This may seem a rather farfetched alternative; how can a star lose mass? The fact is, though, that we know several ways in which a star can lose mass, and a particularly massive star is so likely to lose mass in one of these fashions that we might as well consider the loss inevitable.

Consider the fact that every star will, when its stay on the main sequence comes to an end because its supply of nuclear fuel has dropped below some critical value, expand to a red giant and then collapse.

The more massive the star, the hotter its core by the time of expansion. The combination of larger mass and greater heat produces a larger and larger red giant. Again the more massive the star, the more rapidly it contracts when contraction time comes, since the larger is the gravitational field that powers the contraction.

Suppose we consider a star, then, that is considerably more massive than our Sun and that bloats up to a rather large red planet. The outermost layers of the red giant, which are very far from the denser inner layers, are under a comparatively feeble gravitational pull. When the star contracts, then, the inner layers shoot down rapidly, leaving the outer thinner layers behind. The contracting portion of the star heats up ferociously as the energy of contracting fall is converted into heat. The heat blast strikes the outermost layers, falling inward comparatively slowly, and drives them outward again.

If a star is massive enough, then, and forms a red giant voluminous enough, only the inner portion of it may collapse while the outer portion may be driven away as a turbulent shell of gas. In that case, although the entire star may be above Chandrasekhar's limit, the portion that contracts may be below it and may therefore form a white dwarf.

The result, then, is a white dwarf surrounded by a shell of gas. The white dwarf is very hot as it radiates away the vast energies of the rapid collapse, and the radiation is in the form of ultraviolet light and even more energetic radiations. The shell of gas absorbs this energetic radiation and reradiates it as a soft-colored fluorescence.

What we see from the Earth, then, is a star with a hazy ring around it. It is a shell actually, but the parts of the gas shell toward us in front of the star and away from us on the other side are difficult to see because we are looking through a small thickness of it. On all sides of the star (visible to us), our line of sight is carried through the end of the shell, going through a relatively great thickness of material. The shell therefore looks like a smoke ring. The most remarkable example of this is the Ring Nebula in the constellation of Lyra.

Such nebulas are called *planetary nebulas* because the shell of gas seems to surround the star as though it were in a planetary orbit.

About a thousand planetary nebulas are known, though, of course, many more must exist that we cannot see. Every one of the known planetary nebulas has a small, hot, dense star at the center—probably a white dwarf, though this has actually been demonstrated in only a few of the cases.

If the central stars of planetary nebulas are indeed white dwarfs, they must have been only recently formed, with, as yet, small chance to have radiated much of the heat they have gained from infall. And in point of fact, these are the stars with the hottest known surface temperatures, from at least 20,000°C to, in some cases, well over 100,000°C.

The gas shells we observe seem to have, as nearly as one can tell, a mass equal to a fifth that of our Sun, but larger shells may be possible, too. Some astronomers suggest that a star might lose more than half its mass in the form of a gas shell, and if that is so, a star up to 3.5 times the mass of the Sun can lose enough mass through planetary-nebula formation to allow the collapsing core to drop below Chandrasekhar's limit and form a white dwarf.

Naturally, the gas shell of the planetary nebula, having been driven outward by the energies of the central collapse, is moving outward from the star. The rate of this outward motion can be measured, and figures of 20 to 30 kilometers per second are typical.

As the shell of gas moves farther and farther outward, it stretches out over a larger and larger volume, and its matter gets less and less dense. As it moves farther and farther away from the central star, any portion of the shell receives less and less of the star's radiation and produces less and less fluorescence. The result is that the shell grows dimmer and less visible as it enlarges.

In the typical planetary nebula the shell of gas is from a quarter to half a light-year from the central star, or about 500 times as far from the central star as Pluto is from our Sun.

It has taken perhaps 20,000 to 50,000 years of expansion for the shell to move out to this distance, and this is a short time in the lifetime of white dwarfs. The mere fact that the shell is visible is therefore definite evidence that the white dwarf formed quite recently.

About 100,000 years after white-dwarf formation the shell of gas will have spread outward and thinned out to the point where it is insufficiently luminous to be made out from our vantage point on Earth. It may be, then, that those white dwarfs that lack a shell of gas, lack it only because they have to be well over 100,000 years old.

But the formation of a planetary nebula is not the only way in which a star can lose mass. In fact, there are a number of ways in which we can encounter exploding matter. The big bang may be the largest and most magnificent manifestation of the phenomenon, but there are "little" bangs of one sort or another that are yet enormous enough to be of staggering grandeur.

NOVAS

Anyone watching the cloudless sky night after night with the unaided eye is presented with what seems to be a spectacle of unequaled serenity and changelessness. So much has this changelessness been considered a sign of security in the midst of the turbulent world during our recorded history that any unusual alteration—an eclipse, a shooting star, a comet—was apt to be viewed with fright.

These prominent changes, noticeable to any casual observer, did not affect the stars, however. They were phenomena of our solar system. To an occasional careful observer, though, changes appeared even in the starry universe. Occasionally a new star would appear in the sky where none had been detected before. It was not a shooting star; it remained in place. But it was not a permanent resident, either. Eventually, it would fade out and disappear again.

The greatest of ancient astronomers, Hipparchus of Nicaea (190-120 B.C.), observed such a new star in 134 B.C. and was inspired to prepare the first star map in order that intruders be more easily recognized in the future.

A particularly bright temporary star appeared in November 1572 in the constellation Cassiopeia, and a Danish astronomer, Tycho Brahe (1546-1601), wrote a book about it with the title *De Nova Stella* (which in Latin means "Concerning the New Star"). From this title the expression *nova* came to be applied to temporary stars generally.

In a way the name is a poor one, for the novas are not really new, and they are not truly stars created out of nothing or out of nonstar material, which then return to nothing, or to nonstar material.

Once the telescope was invented in 1608, it became quite clear that there are uncounted millions of stars that are too faint to be seen with the unaided eye. Some of these stars might, for some reason, grow much brighter for a short period of time and then fade again. It could be that a star too dim to be seen without the telescope might brighten to the point of visibility to the unaided eye and then fade to dimness below the level of ordinary vision again. In the days before telescopes it would then seem that the star had come from nowhere and returned to nowhere.

This notion would be much strengthened if some dim star were actually seen to brighten to the level of ordinary vision, but it was not till 1848 that a nova was actually caught in the act. An English astronomer, John Russell Hind (1823-1895) happened to be observing a dim star ordinarily invisible to the unaided eye, when it began to brighten. It reached a peak in the fifth magnitude, by which time it was visible as a dim star to anyone looking at the proper spot in the sky. Then it faded.

Once photography was invented, portions of the sky could be photographed at different times, and comparisons would show whether any star had changed brightness. More novas could be detected in that way; they would not have to be caught in the actual act of brightening. They did not prove to be as uncommon a phenomenon as had earlier been thought. It is now estimated that there might be as many as 30 novas per year on the average in our galaxy.

But what causes a nova?

Whatever it is, it must be something violent. The star that becomes a nova can become thousands or even tens of thousands of times as bright as it was to begin with. What's more, the increase in brightness can take place very rapidly—in as little as a day, or less. After the peak brightness is reached, the decline is never quite as fast as the rise. As the star dims, the rate of further dimming decreases, so that in the end it may take years to return all the way to its prenova state.

The sudden explosive increase in brightness is very likely, then, to be explosive in the literal sense. A close study of the spectrum of novas makes it seem as though shells of gas are emitted by such stars.

Can a nova be a planetary nebula in the making? Can the nova explosion be the last gasp of brightness just before a star collapses to a white dwarf?

Probably not. Before the white dwarf forms, the star should be at the red-giant stage; yet when a star forming a nova has happened to be observed before it became a nova, it did not seem to be a red giant. Besides, the mass of gas ejected by a nova is only about 1/50,000 the mass of our Sun. A planetary nebula does at least tens of thousands of times as well.

Can we expect other kinds of explosions than those forming planetary nebula?

The chances might seem dim at first. After all, most stars seem to be rather stable—as our Sun is, for instance. The gravitational inpull and the temperature outpush are in balance, and a star like our Sun can shine for billions of years without any sudden changes in size or temperature at all. There are sunspots, which slightly cool the Sun, and flares, which slightly heat it, but the changes are very small and are microscopic in comparison to those changes that take place in novas.

Not all stars, however, are as stable as the Sun. There are, for instance, stars whose brightness varies continually, sometimes with rhythmic regularity. This can be because a bright star is partially or wholly eclipsed by a dimmer companion that in its orbit around the bright star passes periodically between it and us.

At other times the variation is due to changes in the star itself.

In 1784 a Dutch English astronomer, John Goodricke (1749-1786)—a deaf-mute who died at the age of 21—noted that the star Delta Cephei (in the constellation Cepheus) varies in brightness. It isn't much of a change: It brightens from magnitude 4.3 to 3.6,* then fades off to 4.3 again, and repeats this over and over. At its brightest Delta Cephei is only twice as bright as it is at its dimmest, and this is not likely to be noticed without a telescope—and in fact, it wasn't.

The nature of the change, however, is a very striking one. The star brightens rather quickly, dims more slowly, brightens rather quickly, dims more slowly, with great regularity, each cycle taking 5.4 days. In the last 200 years, about 700 stars with the same pattern of rather quick brightening and slow dimming over and over have been detected in our galaxy, and all are called *Cepheid variables* in honor of the first to be discovered.

Cepheid variables differ among themselves in the length of their periods. Some have a period as long as 100 days, and some have a period as short as 1 day. (In fact, there are a special group of variable stars, very like the Cepheids, which have periods of 6 to 12 hours and are called *RR Lyrae stars* after the first to be discovered.)

In 1915 the American astronomer Henrietta

---

* As brightness increases, the value of the magnitude decreases.

Swan Leavitt (1868-1921) was able to show that the length of the period depends on the mass and brightness of the star. The more massive and luminous a Cepheid variable is, the longer its period.

Apparently Cepheid variables pulsate, and that is the reason for their changing brightness. The Cepheid variable has reached a stage in its evolution when the balance of gravitation and temperature is no longer smooth. The nuclear fuel supply is perhaps fading to the point where the inner temperature begins to fail. The star therefore begins to collapse, but the very act of collapsing compresses the interior of the star, speeds up the nuclear reactions, and raises the temperature. That forces the substance of the star outward again, and the very act of expanding thins the interior and cools it so that a compression begins again.

The more massive a star, the longer it takes for the swing in and out to go through a complete cycle. This stage is probably short-lived on the astronomical scale, and after a while there will be the final changes leading to expansion to red giant and then collapse.

Can it be that the novas are Cepheid variables in which the pulsation has become extreme? Perhaps as the pulsations continue, they grow wilder and wilder until finally the expansion becomes explosive and the outermost section of the Cepheid is blown off in a process that brightens the star very temporarily, not twofold or threefold, but ten-thousand-fold or more. The loss of the mass might calm the Cepheid variable and return it to a state of sober pulsation again, which, however, may after a time swing outward into explosiveness again. There might be several explosions before the final expansion and collapse.

There are, indeed, stars that have been observed to be *recurrent novas*, which have exploded twice or even three times in the short period of a little over a century in which astronomers have watched such stars closely. What's more, all the Cepheid variables, even the smallest, are considerably more massive than the Sun. They are large, bright stars —just the kind that would have to lose mass if they are to remain within Chandrasekhar's limit and be capable of forming a white dwarf.

It all seems to knit together, but the notion doesn't stand up. A study of stars that go nova, both before they have done so and after they have faded again, shows that they are simply not Cepheid variables. They are not even large stars; they are small and dim, even though they have high surface temperatures.

The combination of smallness and dimness with high surface temperatures suggests white dwarfs; yet white dwarfs are so compact and dense and have such a high surface gravity that they must be very stable. How can they undergo an explosive expansion?

A suggestion, first proposed in 1955 by the Russian American astronomer Otto Struve (1897-1963), that seems to be gaining favor is that every nova is a member of a *close binary*, one of two stars that circle each other at a relatively small distance. One of them, which we will call A, is the larger of the two and therefore reaches the end of its stay on the main sequence before its smaller companion, B, does. As A expands toward the red-giant stage, some of its matter may spill over to B, which is as yet unexpanded. As a result B grows more massive, and A grows less massive. A can then shrink directly to the white-dwarf stage, without passing through a planetary-nebula stage, even though its mass might have been, to begin with, somewhat above Chandrasekhar's limit.

Eventually it is B's turn to leave the main sequence, its lifespan having been shortened by its gain of mass at the expense of A. As B expands toward the red-giant stage, it returns the gift: Some of its matter spills over toward A, which is now a white dwarf.

The surface gravity of A is extremely intense, and the matter it gains undergoes a sudden compression. Since the gained matter will contain some atoms capable of fusion, the compression may eventually produce a very rapid nuclear reaction once enough is gathered and once it is sufficiently compressed. The nuclear reaction releases immense energies that produce a vast flash of light, which accounts for the sudden enormous brightening we see as a nova, and the expulsion of the flashing gas. The nova can recur as additional increments of matter leak over from expanding B.

In this way B can eventually collapse to a white dwarf, even though it had gained enough mass when A expanded to bring it somewhat over Chandrasekhar's limit.

Sirius A and Sirius B would be a good example of this scenario if they were closer together. Unfortunately, their average separation is somewhat greater than that of Uranus and the Sun, so their impingement on each other is limited.

When both were formed, perhaps a quarter of a

billion years ago, the star that is now Sirius B must have been the larger and brighter of the two, with perhaps three times the mass of the Sun; it was shining over Earth (then in the age of dinosaurs) with a brightness equal to that of Venus.

Sirius B did not remain on the main sequence long; it expanded into a red giant and then formed a planetary nebula with perhaps two thirds of its must have been captured by distant Sirius A, shifted outward into invisibility, but some of it must have ben captured by distant Sirius A, whose brightness must have increased and whose life must have shortened as a result. Had Sirius A been considerably closer to Sirius B, it would have picked up much more of the outer layers of Sirius B and might have gained enough mass altogether to have left the main sequence itself soon after Sirius B had. In that case it is possible that Sirius would now be a white-dwarf binary.

As it is, Sirius A will expand to the red-giant stage sometime in the future, and then it will form a planetary nebula. Sirius B is bound to pick up some of the gas shell, possibly enough of it to flare up as a nova. That should be quite spectacular to the descendants of humankind who may then be alive and observing.

We now have two methods by which massive stars can get rid of sufficient mass to fall below Chandrasekhar's limit and form a white dwarf. These two methods—the formation of planetary nebulas and the shifting of matter between the pairs of a close binary—work for stars of only moderate size, up to three times the mass of the Sun. Yet there are still more massive stars. What of them? Let us return to the question of novas.

SUPERNOVAS

Before the days of the telescope the only novas that were sure to be noticed were those which were very bright.

The nova concerning which Tycho Brahe wrote his book, the one that gave the phenomenon its name, was an example of that kind. Tycho's nova at its peak of brightness was 5 to 10 times as bright as the planet Venus, and perhaps 100 times as bright as the brightest stable star, Sirius. Tycho's nova could be seen in daylight, and by night it could even cast a dim shadow that could be seen if the Moon were not in the sky.

Then, in 1604, another bright nova appeared in the constellation Ophiuchus. This one was perhaps only 1/30 as bright as Tycho's nova, but it was still some three times as bright as Sirius. No other nova has since appeared in the sky that has been as spectacular as these two.

There was an earlier case of such a bright nova, however—one that appeared in July 1054 in the constellation Taurus. There are no records of its observation in Europe, which was then just emerging from a "dark age," during which astronomy was temporarily just about nonexistent. We have records, however, from astronomers in China and in Japan.

The 1054 nova, like Tycho's nova, was much brighter than Venus. In fact, the 1054 nova was probably the brighter of the two and could be seen in broad daylight for 23 days. It slowly dimmed after it had reached its peak, but it was nearly two years before the dimming brought it down below the unaided-eye level.

Why were these novas so much brighter than other novas. A logical answer seemed to be that they just happened to be nearer to us than others were and therefore seemed brighter.

In 1885, however, a nova appeared in what was then called the Andromeda Nebula (*nebula is* from the Latin word meaning "cloud.") The "Andromeda Nebula" is a hazy patch of light that astronomers took for granted was a cloud of dust and gas within our own galaxy. The nova, which they assumed just happened to be in the direction of the cloud, wasn't a particularly startling one, for it only reached a maximum brightness of the seventh magnitude and was never bright enough to see without a telescope.

However, as the Andromeda Nebula was observed closely in the years that followed, numerous novas were found within its confines. That many novas could not all be found in one direction; it was too much to ask of coincidence. The notion grew then that the Andromeda Nebula was a distant group of stars, too dim to make out individually, except when one went nova. Eventually, by the 1920s, it was generally agreed that one should speak of the Andromeda Galaxy, which is a galaxy far outside our own, and even larger than our own.

All the novas observed in the Andromeda Galaxy after the 1885 nova were exceedingly dim and were equivalent to the ordinary novas of our own galaxy.

The 1885 nova was something else again. It had to be much brighter than the ordinary novas in either the Andromeda Galaxy or our own. It was so bright that all by itself it had momentarily shone nearly as brightly as all the Andromeda

Galaxy beside. At its peak it was 10 billion times as bright as our Sun and was 100,000 times as bright as an ordinary nova. It was what came to be called a *supernova*, so the 1885 nova was named, in hindsight, S Andromedae, the S standing for supernova.

Once that was settled, it became clear that the bright novas of 1054, 1572, and 1604 were supernovas of our own galaxy.

Supernovas are much less common in occurrence than novas are. Astronomers can see them now and then, here and there, in one distant galaxy or another. Once a supernova comes into existence, it is easy to detect. As soon as a star flares in some galaxy and reaches a peak brightness that makes it as bright as all the rest of the galaxy together, an astronomer knows he has a supernova on his hands. It would seem that there are, on the average, 3 supernovas per millennium per galaxy, as compared with 30,000 ordinary novas. In other words, for every ten thousand novas there is one supernova.

It is difficult to study supernovas in detail when they are located in distant galaxies millions of light-years away. A supernova in our own galaxy would be much more useful, but through a stroke of ill luck not one supernova has been visible to us in our own galaxy since 1604, so that no such close object has ever been investigated telescopically. In fact, in the four centuries since 1604, S Andromedae has been the closest supernova to be observed.

It is clear that the supernova must represent an enormous explosion of a particularly large and massive star. Nothing else could produce radiation at a rate 10 billion times that of the Sun.

What's more, shells of gas are blown off by a supernova, shells that dwarf those produced by planetary nebula in terms of both mass and energy. The best-known example of this is in Taurus, at the site of the great supernova of 1054. There we have a large patch of glowing gas.

This patch was first observed in 1731 by the English astronomer John Bevis (1693-1771). In 1844 the Irish astronomer William Parsons, Lord Rosse (1800-1867), examined it closely with a large telescope he had built and observed that the cloud is filled with irregular filaments that reminded him of the legs of a crab. He called it the Crab Nebula, and that is the name by which it is known to this day.

Close study of the gases of the Crab Nebula shows that they are still moving outward at a rate of about 1,300 kilometers per second. (That rate of outward movement, so much greater than in the case of a planetary nebula, is itself evidence of the incomparable power of the supernova explosion.) Calculating backwards, it seems that all the gas was back at the center just about the time of the 1054 supernova.

Astronomers work backward in other cases. If they find thin wisps of gas in the skies that seem to form part of a shell, they suspect that at one time, in the center of that shell, a supernova had exploded. From the speed of expansion of the shell they can even estimate how long ago the supernova glowed. Some 14 supernova, including the 3 we know, seem to have exploded in our galaxy in the last 20,000 years. If the number in our galaxy was about the same as in other galaxies, there should have been some 60 or 65. The 50 or so we didn't see, however, must have been in distant parts of the galaxy, parts we can't see because of the interposition of the dust clouds that hide much of it from our eyes.

Of the supernova remnants we can detect, the nearest seems to have been one in the constellation of Vela. That supernova, which has given rise to a shell of gas called the Gum Nebula (named for the Australian astronomer Colin S. Gum, who first studied it in detail in the 1950s, and who died in a skiing accident in 1960) has a center only 1,500 light-years away from us, as compared with a distance of 4,500 light-years for the Crab Nebula. The nearest edge of the Gum Nebula is only about 300 light-years from Earth.

The Vela supernova, which gave rise to the Gum Nebula, flashed out some 15,000 years ago, when the Ice Age was just coming to an end. At its peak, it may have been as bright as the full Moon for some days, and we may envy those prehistoric human beings who witnessed that magnificent sight.

What happens to bring about a supernova?

The more massive a star, the higher its internal temperature at every stage in its evolution. A really massive star reaches internal temperatures that smaller stars never do and never can, and we must look for events that happen at those very high temperatures to explain the supernova.

The Chinese American astronomer Hong-Yee Chiu (1932-) has suggested one interesting explanation. The nuclear reactions in the core, he says, give off two kinds of massless particles that travel at the speed of light. One kind is the photon, which is the fundamental particle of light and of lightlike radiation. The other is the *neutrino*.

These two types of particles differ as follows:

The photons are readily absorbed by matter, so they are no sooner formed than they are absorbed. They are then reformed and reabsorbed an indefinite number of times, so they can move at the speed of light only in the tiny, rare intervals between formation and absorption. The result is that it takes about a million years for photons to travel from the core, where they are formed, to the surface, where they escape. The draining of central energy by way of photons is thus very small, and stars, in giving off photons, radiate their energy in a slow, steady way and can therefore last for billions of years.

The neutrinos that are formed do not react with matter at all (or scarcely at all), and once formed in the core, they pass through the outer layers of the star at the speed of light as though nothing were there. It takes about 3 seconds for neutrinos to travel from the core of our Sun to its surface and then fly out into space. It might take 12 seconds for them to travel from the core to the surface of the largest stars on the main sequence. Any energy given off in the form of neutrinos, then, would leak away almost at once.

In ordinary stars, however, the percentage of energy given off in the form of neutrinos is very small, so we usually need consider only the photons.

Chiu suggests, however, that at extremely high temperatures—say, 6 billion degrees—the kinds of nuclear reactions that take place begin to form neutrinos in large quantities. The Sun's internal temperature right now is only about 15,000,000°C, and it will never reach a temperature of 6,000,-000,000°C under any circumstances. Sufficiently massive stars will do so, however, and when the critical point is reached where suddenly vast numbers of neutrinos are being formed, all of them will escape from the star in seconds, taking energy with them and draining the core of the energy required to keep the star expanded against the inward pull of gravity.

As a result the core of the star suddenly cools, perhaps within a matter of minutes, and the star collapses with a precipitousness far beyond anything that can happen in planetary-nebula formation.

In these masive stars, where the core is at the 6-billion-degree mark and where the nuclei have built up to the level of iron, the outer layers are still relatively cool and are still made up of smaller nuclei. As one imagines oneself moving outward from the core, one passes through regions in the star that are less and less evolved, that have more and more of the smaller nuclei that can combine and yield energy and that are at lower and lower temperatures, so that fusion reactions are not yet taking place. In outermost regions of the star there may still even be plenty of hydrogen.

With the sudden, overwhelming implosion of the star the temperature as a whole is raised to enormous values because of the conversion of gravitational energy into heat, and all the nuclear fuel remaining in the star fuses almost at once. This gives rise to the enormous explosion of the supernova and enables the star temporarily to shine as brightly as an entire galaxy of stars.

In the fury of the explosion two things happen. First, many nuclei that are more complex even than iron are formed, for there is a temporary vast energy surplus that makes the formation of such nuclei possible. Second, the explosion drives vast quantities of the star's matter outward as a shell of hot gases containing all the complex atoms that have been formed, up to those with nuclei of as much as five times the size of iron nuclei. Over a period of thousands of years this matter gradually spreads outward, thins, and becomes part of the very thin gases of interstellar space.

Eventually new *second-generation stars* form out of the gases that are in part the remnant of these old stars.

First-generation stars, formed out of the primeval matter of the big bang, are almost entirely hydrogen and helium, and so must their planets be. Nuclei more complex than helium are found only in the core of these stars, and there they are likely to stay—except for supernova explosions.

Second-generation stars, like our own Sun, begin with complex nuclei that the supernovas have spread far and wide, added in small quantities to the hydrogen and helium. The planets of second-generation stars, such as Earth, have those nuclei as well. Life would be impossible without those elements more complex than helium, and all the atoms within our bodies, except for hydrogen, were once at the core of stars that exploded as supernovas.

The enormous explosion of a supernova can drive as much as nine tenths of a star's matter out into space, leaving only a small remnant of itself to collapse and remain collapsed. It isn't hard to assume that a supernova would always leave a remnant that is smaller than Chandrasekhar's limit, so that no matter how large or small a star was, it could always shrink to a white dwarf —quietly if it was less than 1.4 times the Sun's mass, or with an explosion of steadily increasing ferocity as it was more and more over that limit.

Since there are estimated to be three supernovas per millennium per galaxy, and since the universe is estimated to be about 15 billion years old, there may well have been about 45 million supernova explosions in our own galaxy during its history. If all of them gave rise to white dwarfs, they would represent about 1 percent of the total number of white dwarfs estimated to exist in our galaxy.

That seems reasonable. We can suppose that only the very massive stars undergo a supernova explosion, while smaller ones reach the white dwarf by way of planetary-nebula explosions or still quieter contractions. And there are more small stars than large ones, so there should be many more white dwarfs than there have been supernova explosions. (It should be remembered, however, that even the "small stars" mentioned in this connection are not much smaller than our Sun. None of the really small stars that make up the large majority have yet lived long enough to reach the point of expansion and collapse, not even if they were born at the moment of the big bang.)

So it might seem that we have a clear picture of the end of stars, and that end is always the white dwarf cooling to the black dwarf. Yet some astronomers were not satisfied—.

NEUTRON STARS

BEYOND THE WHITE DWARF

INDIVIDUAL STARS have been detected with as much as 50, possibly 70, times the mass of our Sun. When such a star goes, it will go with an unexampled crash. What's more, when it does go, it will have to get rid of 97 or 98 percent of its mass in order that what is left over will be only 1.4 times the Sun's mass and can safely collapse to a white dwarf.

That may happen, of course, but what if it doesn't? Astronomers know that supernovas get rid of a lot of mass, but there is nothing in the process, as far as they know, that says a supernova *must* get rid of enough mass to leave a contracting body below Chandrasekhar's limit. What if, after a supernova explosion, what is left of the star has a mass twice that of the Sun and that this two-Sun mass collapses. The electronic fluid will form and contract—and contract—

and smash! The inward pull of gravity will simply be too intense to be balanced by the electronic fluid at its most compressed.

The electrons will then be driven inward to densities at which they cannot really exist. Within the electronic fluid protons and neutrons had been moving about freely; now the electrons will combine with the protons to form additional neutrons. The electrons and protons are present in any piece of matter, whether a dust fragment or a star, in just about equal numbers, so the result of the union is that the collapsing star will consist just about entirely of neutrons.

These neutrons will be driven together by the gravitational collapse until they are in virtual contact. Then, and only then, will the collapse be halted. The nuclear force, which governs the interaction of massive particles, keeps the neutrons from pushing any closer together. Now it is no longer gravitational force being balanced by electromagnetic force as it is in planets, in ordinary stars, and even in white dwarfs. It is gravitational force balanced by the nuclear force, which is much stronger than the electromagnetic force.

A star consisting of neutrons in contact is called a *neutron star*. It is composed of a neutronic fluid that is sometimes referred to as *neutronium*. In a sense an atomic nucleus is made up of neutronium, and, in reverse, a neutron star is like a giant nucleus. Neutronium is incredibly dense; it reaches a peak of something like 1,000,000,000,000,000, or 10^{15}, times as dense as ordinary matter.

If a sphere of ordinary matter were converted into a sphere of neutronium, its diameter would shrink to 1/100,000 the orginal without loss of mass. Thus, the Earth, which is 12,740 kilometers in diameter, would, if it were suddenly turned into neutronium, be a sphere about 0.127 kilometers in diameter. A sphere that size is only 1.5 city blocks across, but it would contain all the mass of the Earth.

Similarly, if the Sun, which is 1,400,000 kilometers in diameter, were converted into neutronium, it would become a sphere only 14 kilometers across. It would have the volume of a small asteroid, but it would have all the mass of the Sun.

It isn't safe, as we shall see, to imagine neutron stars much more massive than the Sun, but just to get a clear picture we can imagine the most massive known star somehow converted into neutronium without the loss of any of its mass. It would be a sphere only 50 or 60 kilometers across.

Even the cosmic egg has, at times, been imagined as a gigantic ball of neutronium containing all the mass of the universe—a "neutron universe," so to speak. It would be 300,000,000 kilometers across. If such a cosmic egg were put in the place of our Sun, it would reach out only to the asteroid belt, yet it would contain all the mass of the 100,000,000,000 stars of our galaxy and of all the stars of 100,000,000,000 other galaxies.

Nor do we have to imagine only masses above Chandrasekhar's limit as forming neutron stars. When a supernova explodes, the collapse of that portion of the star that is not blown away is so sudden that it comes smashing down on the electronic fluid with incredible speed. It is, then, not the sheer mass so much as the rapid infall that breaks through the electronic-fluid barrier. Once the electronic fluid is smashed, that smash is irreversible. The electronic fluid cannot reconstitute itself. As a result a neutron star with as little as a fifth the mass of our Sun may form, with a diameter of only 8.2 kilometers.

The likelihood that the force of supernova collapse can smash the electronic fluid even when the collapsing mass is under Chandrasekhar's limit makes it look as though supernovas are bound to form neutron stars. White dwarfs will form only when stars too small to explode as a supernova reach their cycle of expansion and contraction with nothing worse than a planetary nebula developed.

In 1934 the Swiss-American astronomer Fritz Zwicky (1898-1974) and the German-American astronomer Walter Baade (1893-1960) were the first to speculate about the possible formation and existence of neutron stars. A few years later the American physicist J. Robert Oppenheimer (1904-1967) and a student of his, George M. Volkoff, worked out the theory in detail.

But then came World War II, which preoccupied scientists to the exclusion of almost everything else, Oppenheimer, for instance, headed the team that developed the nuclear bomb.

Even discounting the pressures of war work, however, interest in neutron stars was not very widespread among astronomers. After all, the matter seemed hopelessly theoretical. An astronomer might work out exactly what could happen in a supernova explosion. He might calculate the manner in which matter might be blown away, what the speed of collapse might be, at what point the electronic fluid would be smashed, and how the neutronium might form—yet it would all just remain figures on paper.

How could one prove that the theory was correct and that neutron stars exist? Was it reasonable to suppose that an object that is 8 to 15 kilometers across and, surely, light-years away could be seen?

Even if a neutron star were as intensely bright as the brightest star, its tiny, tiny surface would deliver only the feeblest spark. Even if the biggest and best telescope were focused in its direction, it would show up, at best, as a very dim, dim star. How could one possibly tell it was a neutron star that happened to be close enough to be made out, rather than an ordinary star that looked that dim only because it was extremely far away?

Then, why bother about neutron stars?

Well, as long as the only important way in which astronomers could study the sky was by observing the light given off by objects in it, it was useless to bother. As the twentieth century progressed, however, astronomers became more and more aware of cosmic radiation other than that of light, and eventually the problem of detecting a neutron star did not seem so impossible after all.

BEYOND LIGHT

In 1911 the Austrian-American physicist Victor Francis Hess (1883-1964) was able to show that certain very energetic forms of radiation reach Earth from space, and they were therefore called *cosmic rays*.

The cosmic rays are composed of very fast, electrically charged atomic nuclei that very likely originated in the millions of supernovas that have exploded in our galaxy during its lifetime. Because the cosmic-ray particles are electrically charged, however, they curve in their paths in response to the various magnetic fields associated with stars and with the Galaxy as a whole. They end up reaching us from all directions, and there is no way we can tell from what specific direction a specific cosmic-ray particle began its travels. While cosmic rays continue to be interesting to astronomers, they cannot be used to give us information about particular stars.

In 1931 the American radio engineer Karl Guthe Jansky (1905-1950) discovered that there are *microwaves* reaching us from the sky. Microwaves are lightlike radiations without electric charge, so they travel in straight lines, unaffected by magnetic fields. Microwaves, as the name implies, are made up of waves, as light is, but microwaves are about a million times as long as lightwaves are.

Despite this, the *micro* of microwaves is from a Greek word meaning "small" because microwaves belong to a group of radiations called *radio waves*, and microwaves are the smallest of that particular group. (Microwaves are often referred to as radio waves, by the way.)

Because the microwaves are so long, compared with light waves, they have less energy and are less easily detected. Furthermore, the accuracy with which a wave source can be pinpointed decreases as the wavelength increases, all other things being equal. It was therefore much harder to work out the point of origin of microwaves than of light. For quite a while, therefore, little was done with microwaves.

The existence of microwaves reaching us from the sky, made it clear that stars radiate in all wavelengths. It so happens that the short wavelengths of ordinary light and the long wavelengths of microwaves happen to be able to get through our atmosphere, while other wavelengths cannot. The atmosphere is more or less opaque for one reason or another to the wavelengths shorter than those of visible light, longer than those of microwaves, or intermediate between the two.

In the early 1950s rockets began to be sent beyond the atmosphere into space to observe and measure those ranges of wavelengths blocked by the atmosphere. Rockets could only remain beyond the atmosphere for short periods of time at first before returning and plunging back to Earth.

Starting in 1957, however, first the Soviet Union, then the United States began to place satellites in orbit around the Earth beyond the atmosphere. They could remain beyond the atmosphere for indefinite periods, and they could carry instruments that could detect the full range of radiation coming from the sky. With the proper instruments they could detect *ultraviolet light*, which has wavelengths shorter than those of visible light; *X-rays*, with still shorter wavelengths; and even *gamma rays* with even shorter wavelengths.

This roused hope, for violent events involve higher temperatures and therefore more energetic radiation. Any star can radiate light, but only violent stars—therefore interesting stars—will radiate X rays, for instance.

As an example, our own Sun emits X rays from its thin outer atmosphere, the *corona*. This is because the heat pouring out of the Sun is absorbed by the thinly spread-out atoms of the corona, and each atom is therefore raised to a temperature of a million degrees or more. (The total heat of the corona is, however, not very great, since although the individual atoms are so hot, there are so few of them.)

Because the Sun is so close, it is the most important emitter of X rays in the sky, but if it were at the distance of even the nearer stars, its X-ray radiation would be so diluted by distance it could not be detected. Sirius, for instance, is considerably larger and hotter than our Sun, and therefore undoubtedly emits X rays with several times the intensity that the Sun does. Yet Sirius is at a distance of nearly nine light-years, and its X rays cannot be detected.

If X rays could be detected at star distances, it would indicate violence indeed, but at first astronomers didn't think such detections could be carried through. Their assumption in the early 1960s was that the Sun was the only source of detectable X rays in the sky. Nevertheless, there was some interest in studying the night sky, since it was possible that solar X rays might be reflected from the Moon and that this could give us some information about the Moon's surface. (This was before astronauts landed on the Moon.)

In 1963 under the guidance of the American astronomer Herbert Friedman (1916-) investigations beyond the atmosphere were conducted for the detection of X rays coming from the Moon. Such X rays were not detected, but X rays were found, quite surprisingly, from other directions. In the years since, some satellites have been sent up for no other purpose but to map the sky for X-ray sources, and hundreds of such sources have been located.

This gave the universe an entirely new aspect. An X-ray source that could be detected at the distance of the stars and even, in many cases, at the distance of the other galaxies, must mark events quite out of the common.

For one thing, the existence of such X-ray sources gave rise to hopes that neutron stars might be detected. When a neutron star is formed, it is, in a way, like an exposed core of a star and possesses at its surface the temperature of a stellar interior. Theoretical considerations made it seem that the surface of a neutron star would glow at a temperature of 10,000,-000°C.*

A neutron star with a surface at that temperature would radiate chiefly in the X-ray region. Consequently, the question arose as to whether some of the X-ray sources in the sky might not originate in neutron stars.

That wasn't the only possibility, of course. X rays might originate from the very hot gases pushed out by supernovas, for instance, just as they originate from the Sun's corona.

These two possibilities can be distinguished as follows: A neutron star would be a tiny point in the sky, while a region of gases would be a dis-

---

* If the cosmic egg were a gigantic neutron star, its surface temperature would probably be 1,000,000,000,000°C at least, and it would radiate gamma rays.

tinct smear. A lot would depend, then, on whether the X rays seemed to emerge from a single point or from an area.

A prime suspect was the Crab Nebula. It is the remains of a tremendous supernova, and there *might* be a neutron star somewhere in the center of all those gases. And, of course, the gases are there, and they are clearly in energetic turmoil. X rays might come from the suspected neutron star if one were there, or from the gases, or from both.

In 1964 the Moon was slated to move in front of the Crab Nebula. As the Moon advanced, it would cut off the X-ray emission. If the X rays were coming only from the pointlike neutron star, they would remain at full intensity as the Moon advanced and then sink suddenly at zero. If the X rays were coming from the gas, they would decline smoothly in intensity. If the X rays were coming from both, they would decline smoothly at first, then experience a sudden drop, then decline further as smoothly as at the start.

A rocket was sent up at the appropriate time to measure the X-ray intensity from the Crab Nebula, and the measurement fell off gradually, more or less, as the Moon advanced. The X rays seemed to be coming from the turbulent gas, and hopes for the detection of a neutron star withered.

PULSARS

Meanwhile, however, astronomers had begun working with microwaves, and the science of *radio astronomy* had been rapidly developed to a high pitch of complexity and efficiency. Astronomers learned to use complex arrays of detecting devices (*radio telescopes*) in such a way as to be able to pinpoint microwave sources with great accuracy and to work out their properties in great detail.

In the early 1960s, for instance, radio astronomers became aware that some microwave sources change intensity rather rapidly, as though they were twinkling. They began to design radio telescopes that were specially adapted to catch the rapid changes. One such radio telescope was devised at Cambridge University Observatory by Anthony Hewish (1924-) and consisted of 2,048 separate receiving devices spread out over an area of 18,000 square meters.

In July 1967 the new radio telescope was set to scanning the heavens, and within a month a young graduate student, Jocelyn Bell, was receiving bursts of microwaves from a place midway between the stars Vega and Altair—very rapid bursts, too. At first, she thought she was detecting interference with the radio telescope's workings from electrical devices in the neighborhood. However, she discovered that the sources of the microwave bursts move regularly from night to night across the sky in time with the stars. Something outside the Earth had to be responsible for it, and she reported the results to Hewish.

By the end of November the phenomenon could be studied in detail. Hewish had expected rapid fluctuation, but not that rapid. Each burst of microwaves lasted only 1/20 second, and the bursts came at intervals of 1 1/3 seconds. They came, indeed, with remarkable regularity. They came every 1.33730109 seconds.

The new instrument picked up these bursts of microwaves easily, for the individual bursts were energetic enough to detect without trouble. Ordinary radio telescopes, however, had not been designed to catch these very brief bursts; they had detected only an average microwave intensity including the dead period between bursts. This average is only 3.7 percent of the burst maximums, and this had gone unnoticed.

The question was: What does this phenomenon represent? Since the microwave source seems to be a mere point in the sky, Hewish thought it might represent some kind of star. Since the microwaves emerge in short pulses, he thought of it as a kind of *pulsating star*. This was shortened almost at once to *pulsar*, and it was by that name that the new object came to be known.

Hewish searched for others among the long charts of previous observations by his instruments and found three more pulsars. He checked the evidence, and then on February 9, 1968, he announced the discovery to the world.

Other astronomers began to search avidly, and more pulsars were quickly discovered. By 1975 100 pulsars were known, and there may be as many as 100,000 in our galaxy altogether.

Two thirds of the pulsars that have been located are to be found in those directions where the stars of our galaxy are thickest. That is a good sign that pulsars generally are part of our own galaxy. (There is no reason to suppose they don't exist in other galaxies, too, but at the great distances of other galaxies they are probably too faint to detect.) The nearest known pulsar may be as close as 300 light-years or so.

All the pulsars are characterized by extreme

regularity of pulsation, but of course the exact period varies from pulsar to pulsar. The one with the longest period has one of 3.75491 seconds.

The pulsar with the shortest period so far known was discovered in October 1968 by astronomers at Green Bank, West Virginia. It happens to be in the Crab Nebula (making the first clear link between pulsars and supernovas) and proved to have a period of only 0.033099 seconds. It is pulsing 30 times a second, or 113 times as rapidly as the pulsar with the longest period known.

But what can produce such short flashes in such a fantastically regular fashion?

So stunned were Hewish and his fellow astronomers at the first pulsars that they wondered if it were possible they might be signals from some intelligent life-forms far out in space. Indeed, among themselves they referred to the matter as *LGM* before the word *pulsar* came into use— LGM standing for "little green men."

This notion didn't last long, however. To produce the pulses would require 10 billion times the total quantity of power humankind could produce. It didn't seem likely that so much power would be wasted just to send out very regular signals that carried virtually no information. Besides, as more and more pulsars were detected, it seemed quite unlikely that so many different life-forms would all be zeroing in their signals on us. The theory was quickly dropped.

But something must be producing them; some astronomical body must be undergoing a steady periodic change—a revolution around some other body, a rotation about its own axis, a pulsation— at intervals rapid enough to produce the pulses.

To force such rapid changes with the release of so much energy would require an enormously intense gravitational field. Astronomers knew of nothing else that would work. Instantly white dwarfs came to mind.

Theoreticians got busy at once, but try as they might, there seemed no way of allowing one white dwarf to circle another, or to rotate on its axis, or to pulse, with a period short enough to account for pulsars. Small and gravitationally intense white dwarfs might exist, but they could not be small enough nor could their gravitational fields be intense enough for the task. They would actually break up and tear apart if they were to revolve, rotate, or pulse in periods of less than four seconds.

Something smaller and denser than a white dwarf was required, and the Austrian-born astronomer Thomas Gold (1920-) suggested that pulsars are the neutron stars that Oppenheimer had played with theoretically. Gold pointed out that a neutron star is small enough and dense enough to be able to rotate about its axis in four seconds or less.

What's more, a neutron star should have a magnetic field just as an ordinary star does, but that magnetic field should be compressed and concentrated as the matter of the neutron star is. For that reason, a neutron star's magnetic field is enormously more intense than the fields about ordinary stars. The neutron star as it whirls on its axis gives off electrons, but they are trapped by the magnetic field and are able to escape only at the magnetic poles, which are at opposite sides of the star.

The magnetic poles need not be at the actual

rotational poles. (They aren't in the case of our own Earth, for instance.) Each magnetic pole might sweep around the rotational pole in seconds or in a fraction of a second and spray out electrons as it does so (just as a rotating water sprinkler jets out water). As the electrons are thrown off, they curve in response to the neutron star's magnetic field and gravitational field. Losing energy, they may not escape altogether, but the energy they lose is in the form of microwaves.

Every neutron star thus shoots out two jets of microwaves from opposite sides of its tiny globe. If a neutron star happens to move one of those jets across our line of sight as it rotates, Earth will get a very brief pulse of microwaves at each rotation. Some astronomers estimate that only one neutron star out of a hundred would just happen to send microwaves in our direction, so of the possibly 100,000 in our galaxy we might never be able to detect more than 1,000.

Gold went on to point out that if his theory was correct, the neutron star is leaking energy at the magnetic poles, and its rate of rotation must be slowing down. This means that the faster the period of a pulsar, the younger it is likely to be and the more rapidly it may be losing energy and slowing down.

The most rapid pulsar known, and the one with the most energetic pulses, is the Crab Nebula pulsar, and it might well be the youngest we happen to have observed so far, since the supernova explosion that might have left that neutron star behind took place only 900 years ago. At the very moment of its formation the Crab Nebula pulsar might have been rotating on its axis 1,000 times a second, but it would have lost energy quickly; in the first 900 years of its existence over 97 percent of its energy has bled away until it is now rotating only 30 times a second. And it should still be slowing, though, of course, more and more slowly.

The period of the Crab Nebula was studied carefully, and the pulsar was indeed found to be slowing down, just as Gold had predicted. The period is increasing by 36.48 billionths of a second each day, and at that rate it will have doubled in 1,200 years. The same phenomenon has been discovered in other pulsars whose periods are slower than that of the Crab Nebula pulsar and whose rate of slowing is also slower. The first pulsar discovered, now called CP1919, has a period 40 times as long as that of the Crab Nebula pulsar, and it is slowing at a rate that will double its period only after 16 million years. As a pulsar slows, its pulses become less energetic. By the time the period has passed four seconds in length, the pulsar becomes too weak to be detectable. Pulsars probably endure as detectable objects for tens of millions of years, however.

As a result of these studies of the slowing of the pulses astronomers are now pretty well satisfied that the pulsars are neutron stars.

Sometimes a pulsar will suddenly speed up its period very slightly, then resume the slowing trend. This was first detected in February 1969, when the period of the pulsar Vela X-1 (found amid the debris of the supernova that blazed up 15,000 years ago) was found to alter suddenly.

The sudden shift was called, slangily, a *glitch*, from a Yiddish word meaning "to slip," and glitch has entered the scientific vocabulary in consequence.

Some astronomers suspect glitches may be the result of a *starquake*, a shifting of mass distribution within the neutron star that will result in its shrinking its diameter by a centimeter or less. Or perhaps it might be the result of a sizable meteor plunging into the neutron star and adding its own momentum to that of the star.

There is, of course, no reason why the electrons emerging from a neutron star should lose energy only as microwaves. They should produce waves all along the spectrum. It should, for instance, emit X rays, too, and the Crab Nebula neutron star does, indeed, emit them. About 10 to 15 percent of all the X rays the Crab Nebula produces is from its neutron star; it was the other 85 percent or more that comes from the turbulent gases that obscured this fact and disheartened those astronomers who hunted for a neutron star there in 1964.

A neutron star should produce flashes of visible light, too. In January 1969 it was noted that the light of a dim sixteenth-magnitude star within the Crab Nebula does flash on and off in precise time with the microwave pulses. The flashes and the period between them are so short that special equipment was required to catch them. Under ordinary observation the star seems to have a steady light. The Crab Nebula neutron star was the first *optical pulsar* discovered, the first visible neutron star—up to now the only one.

PROPERTIES OF NEUTRON STARS

Astronomers speculate about the detailed composition of neutron stars. At the very surface there may be a thin layer of normal matter, mostly iron. There may even be a gaseous iron atmosphere, perhaps half a centimeter thick. There are also charged particles such as electrons and atomic nuclei that are bound to the neutron star's superintense magnetic field. It is these, the electrons particularly, that are sprayed out at the magnetic poles and that produce the pulses of radiation that are detected on Earth.

Below that outermost shell of normal matter, are well-packed iron nuclei, bearing characteristics we would think of as "solid," even though this crust is at a temperature of millions of degrees. The outer edge of the crust has a density of only 100,000 g/cm^3, but this rapidly increases with depth.

It is this solid surface, with strength nearly a billion billion times that of steel and with "mountains" possibly a centimeter high, that readjusts itself every now and again to settle down into a more compact form producing the glitches that slightly decrease the period of rotation.

Below the crust, as density increases further, nuclei cannot maintain their integrity, and the material becomes a mass of neutrons. Near the core there may be a sea of still more massive particles called *hyperons*.*

One important property of the neutron star is

---

* Hyperons can be produced in laboratories, but under earthly conditions they break down in less than a billionth of a second.

its mass. In 1975 the mass of a neutron star was determined for the first time. The neutron star in question, Vela X-1, turned out to have a mass 1.5 times that of the Sun. This was interesting, for the mass was slightly over Chandrasekhar's limit. No white dwarf could have been that massive (although we must remember that neutron stars with masses considerably below Chandrasekhar's limit are also possible in theory).

The mass of Vela X-1 was capable of being determined because that neutron star is part of a binary. Its companion is a massive star of the main sequence, one with 30 times the mass of our Sun. Undoubtedly binaries, if massive enough, can shift matter back and forth as each expands, and end by forming a pair of neutron stars, just as less massive binaries can in this fashion produce a pair of white dwarfs.

Vela X-1 must orginally have been the brighter of the pair, and 15,000 years ago, when it became a supernova, the companion star may have captured as much as a thousandth of the matter blown off by the explosion, gaining considerably in mass and brightness in consequence and, of course, shortening its own life on the main sequence. Eventually, in a million years or less, the companion of Vela X-1 will go supernova in its own right, and there may then be two neutron stars rotating about a common center of gravity. The fact that a neutron star can form part of a binary, as Vela X-1 does, shows that when one star of a pair goes supernova, the other star can survive.

The shift of matter from one star to another as first one and then the other expands results in the conversion of gravitational energy to radiation, especially where a white dwarf or a neutron star, with a very intense gravitational field, is involved. Up to 40 percent of the mass of matter can be converted into energy in this way—more than 100 times the amount of mass that can be converted to energy by way of nuclear fusion. This is another point that helps explain the brightness of novas and supernovas.

Next, let's consider some of the gravitational properties of a neutron star, taking as our average specimen one that has exactly the mass of our Sun but a diameter only 1/100,000 as great. Such a neutron star must have a diameter of 14 kilometers and an average density of 1,400,-000,000,000,000 g/cm³.

If we consider the Sun first, its surface gravity is equal to 28 times that of Earth's surface gravity. Thus, a person who weighs 70 kilograms on the surface of the Earth would weigh on the surface of the Sun (assuming the Sun had a surface in the earthly sense and that a person could survive the experience) just under 2,000 kilograms.

Now, if we imagine a body of a given mass being compressed smaller and smaller, any object on its surface comes closer and closer to the center. By Newton's law of gravitation the surface gravity (assuming the mass doesn't change) changes inversely * as the square of the diameter. Thus, if you compress a star so that it has only 1/2 its original diameter, the surface gravity is 2 × 2, or 4 times the original. If it is compressed to 1/6 its original diameter, then the

---

* By *inversely* we mean that surface gravity and diameter change in opposite direction. As diameter decreases, surface gravity increases; as diameter increases, surface gravity decreases.

surface gravity is 6 × 6, or 36 times the original; and so on.

Sirius B, with a diameter 1/30 that of the Sun and a mass just about equal to it, must have a surface gravity 30 × 30, or 900 times that of the Sun. Our mythical 70-kilogram person, who can survive any experience, would on the surface of Sirius B weigh 1,800,000 kilograms.

A neutron star with the mass of the Sun and a diameter of 14 kilometers (1/100,000 times that of the Sun) must have a surface gravity 100,000 × 100,000, or 10,000,000,000 times that of the Sun. Our 70-kilogram person would weigh 20 trillion kilograms.

And what about rotational periods?

Our Earth, with a circumference of 40,000 kilometers, rotates on its axis in one day. That means that a point on Earth's equator, which marks out a larger circle in that one day of rotation than any other point not on the equator does, is traveling around Earth's axis at a constant speed of just about 0.5 kilometers per second. This speed decreases steadily as one moves farther and farther away from the equator, either north and south, until it is zero at the poles.

A rotational speed sets up a centrifugal effect that tends to counter the pull of gravity. This centrifugal effect increases with speed of rotation, so it is zero at the poles and increases as one approaches the equator until it is a maximum at the equator. The centrifugal effect tends to pull material away from the axis, and do it most strongly at the equator, so that we can say that the Earth has an *equatorial bulge*. It isn't much. The equatorial diameter (the distance from one point on the equator to the opposite point, through the Earth's center) is 43 kilometers longer than the polar diameter (from pole to pole). The equatorial diameter of Earth is roughly 1/300 longer than the polar diameter, and that is a measure of Earth's *oblateness*.

Consider Jupiter, on the other hand. Jupiter, the largest planet, has an equatorial circumference of 449,000 kilometers and rotates in 9.85 hours. A point on Jupiter's equator therefore moves at a speed of 12.7 kilometers per second, just over 25 times as fast as a point on Earth's equator.

Despite Jupiter's greater gravity, this enormous speed of rotation, combined with the fact that Jupiter's substance is composed of lighter elements much less compactly packed than Earth's substance is, results in a larger oblateness for Jupiter. Jupiter's equatorial diameter is 8,700 kilometers longer than its polar diameter. Its oblateness is fully 1/16.*

The Sun by comparison has a circumference of 4,363,000 kilometers and rotates on its axis in 25.04 days. A point on its equator therefore moves at a speed of just about 2 kilometers per second. This is 4 times the speed of a point on Earth's equator, but it is only 1/6 the speed of a point on Jupiter's equator. The combination of relatively slow rotational speed of the Sun, and its huge surface gravity is such that no oblateness can be measured. As far as we can tell, the Sun is a perfect sphere.

We don't know what the period of rotation is for Sirius B, or for any white dwarf, but we

---

* Saturn is a bit smaller than Jupiter and does not rotate quite as fast, but its gravitational field is also smaller, and it is even more oblate than Jupiter is.

know that a typical neutron star will rotate in about 1 second, judging from the period of pulsation of the pulsars. If our 14-kilometer-across neutron star rotates about its axis in 1 second, then a point on its equator will be moving at a speed of about 44 kilometers per second.

This is 3.5 times as fast as a point on Jupiter's equator, 21.8 times as fast as a point on the Sun's equator, and 95 times as fast as a point on the Earth's equator. Nevertheless, considering the neutron star's vastly intense gravitational field, we can be quite certain that its rotational speed, fast though it might be by solar-system standards, simply can't even approach being able to lift any material against gravity through a centrifugal effect. The neutron star must be a perfect sphere no matter what. We can be almost as confident that the white dwarf must be a perfect sphere also.

If centrifugal force is not likely to lift the substance of white dwarfs and neutron stars a measurable distance against gravity, we can imagine that the escape velocity from such bodies must be high indeed, and we would be right.

Escape velocity varies inversely as the square root of the diameter (assuming no change in mass). Thus, if you decrease a star to 1/36 times its original diameter, then the escape velocity increases by 6 times (since 6 is the square root of 36).

Working on this basis, you can see that Sirius B, with a mass equal to that of the Sun and a diameter 1/30 of the Sun, must have an escape velocity 5.5 times that of the Sun. Since the escape velocity from the Sun's surface is 617 km/sec, that from the surface of Sirius B must be 3,400 km/sec.

On the other hand, our neutron star, with its mass equal to the Sun but with a diameter only 1/100,000 as great, must have an escape velocity at its surface that is greater than that of the Sun by a factor equal to the square root of 100,000, or 316. The escape velocity from the neutron star must be equal to 617 × 316, or just about 200,000 km/sec.

These figures on escape velocity are particularly important to us because they are another milestone on the road to the black hole. Let us therefore present them in tabular form in Table 12.

TABLE 12—Escape Velocities

Object	Escape Velocity	
	Kilometers per second	*Fraction of speed of light*
Earth	11.2	0.0000373
Jupiter	60.5	0.00020
Sun	617	0.0020
Sirius B	3,400	0.011
Neutron star	200,000	0.67

For objects of ordinary matter escape velocities are tiny fractions of the velocity of light. Even for the Sun the escape velocity is only 1/500 the velocity of light. In the case of the white dwarf the escape velocity is 1/100 the velocity of light, and light itself loses a measurable amount of energy in leaving. It was by this loss of energy and the consequent small red shift in the light of Sirius B that Adams was able to check its dense nature.

A neutron star is likely to have an escape velocity equal to 2/3 that of light, and the Einstein shift would be much greater. We may get X radiation from a neutron star, but if it were not for the star's intense gravitational effect, X rays we receive would have far shorter waves than they in fact have. And as for the long-wave radiation we get, the visible light waves and the much longer microwaves, much of that too would not exist were it not for the wave-lengthening effects of the neutron star's gravitational field.

TIDAL EFFECTS

There is another gravitational effect that we can neglect on Earth's surface but that becomes of overwhelming importance in the neighborhood of a neutron. This is the *tidal effect*.

The strength of the gravitational attraction between two particular objects of given mass depends on the distance between their centers. For instance, when you are standing on Earth's surface, the strength of Earth's gravitational pull on you depends on your distance from Earth's center.

Not all of you, however, is at the same distance from Earth's center. Your feet are nearly two meters closer to the Earth's center than your head is. That means that your feet are more strongly attracted to the Earth than your head is because gravitational attraction increases as distance decreases. This difference in the gravitational attraction between two ends of an object is the tidal effect.

Under ordinary circumstances tidal effects aren't great. Let us consider a person (a rather large one) who is two meters tall and who weighs 90 kilograms. If he is standing on the Earth at sea level in the United States, the soles of his feet are about 6,370,000 meters from the center of the Earth. Let us say they are at exactly that distance. In that case the top of his head is about 6,370,002 meters from the center of the Earth.

The gravitational pull at the top of his head is $(6,370,000/6,370,002)^2$ times the gravitational pull at the soles of his feet. This means that the pull on his feet is about 1.0000008 times the pull on his head. This is the equivalent of saying that he is on a rack with the top of his head and the bottom of his feet being stretched apart by the pull of a weight of 0.000071 kilograms, which is the equivalent of about four drops of water. This sort of pull is too small to be felt, and that is why we are not aware of tidal effects produced by the Earth on our body.

The tidal effect is greater if an object subjected to a gravitational field is larger, so that there is a larger drop in the force exerted upon the object between one end and the other. Instead of a person, let's choose the Moon.

The Moon has a diameter of 3,475 kilometers, and its center is at an average distance of 384,-321 kilometers from the Earth's center. If we imagine the Moon to be always at that distance (there is actually a slight variation in and out during the month but not a very large one), then the part of its surface directly facing Earth would be 382,584 kilometers from Earth's center, and the part of its surface directly away from Earth would be 386,058 kilometers from Earth's center.

The gravitational pull on the near side of the Moon, because it is nearer, would under these circumstances be 1.018 times that on the far side.

The total force of Earth's gravitational pull on the Moon (what we would imagine its weight to be if it were resting on a platform attracted to the center of the Earth and 384,321 kilometers high) would be 20,000,000,000,000,000,-000 kilograms.

If the Moon were all at the distance of its nearest surface, then it would weigh 800,000,-000,000,000,000 kilograms more than if it were all at the distance of the farthest part of the surface. You can imagine, then, the Moon being stretched in the direction toward and away from the Earth by that amount of pull; 800 million trillion kilograms is no mean pull, and the Moon shows a small bulge in that direction. The diameter pointing toward and away from the Earth is slightly longer than the diameter at right angles to it.

It works the other way around, too. The Moon pulls on the Earth, and it pulls more strongly on the side of the Earth nearest itself than on the part farthest from itself. Since the Earth has a greater diameter than the Moon does, there is a longer distance over which the gravitational pull can decrease, and that makes for an increase in the tidal effect. The Moon is a smaller body than the Earth is and produces a smaller gravitational pull altogether, and that makes for a decrease in the tidal effect.

The decrease wins out. The smaller gravitational field of the Moon is more important a factor than the greater diameter of the Earth. If the gravitational effect were all important, the Moon's tidal effect on Earth would be 1/81 the Earth's tidal effect on the Moon. Earth's greater diameter slightly compensates, and in fact the Moon's tidal effect on the Earth is 1/70 the Earth's tidal effect on the Moon.

The Earth is stretched in the direction of the Moon by a perceptible amount. The solid ball of the Earth stretches by about a third of a meter. The water of the ocean gives more easily and stretches by just over a meter.

There is therefore a bulge in the ocean (and a lesser one in the solid crust) on the side facing the Moon, and another on the opposite side of the Earth, away from the Moon. As the Earth rotates, the land surfaces move into the bulge and out again, then into the other bulge and out again. As a result the ocean creeps up the shore and down again twice a day (in a way strongly affected by the shape of the shoreline and other factors we need not go into in this book). This two-a-day ocean movement is referred to as the tides, and that is why the phenomenon is referred to as the tidal effect.

The tidal effects of bodies such as the Earth and Moon are not really very large compared with the total gravitational force, but they mount up with time. As the Earth turns through the bulges, the friction of the water against the bottom of the shallower portions of the ocean, converts some of the rotational energy into heat. As a result the Earth is slowly decreasing its rate of rotation and slowly increasing the length of its day. The day becomes 1 second longer every 100,000 years. That doesn't sound like much, but if this has been a steady rate of decrease, the Earth rotated in only 12.7 hours when it was first formed.

The earth can't lose *angular momentum* (something that involves its rate of turning) without that being gained elsewhere in the Earth-Moon system. The Moon gains it and is slowly moving farther away from the Earth as a result, since that is a movement that increases its angular momentum.

The Earth's tidal effect on the Moon has slowed the Moon's rotation to the point where it faces one side to the Earth at all times.

Like gravitation as a whole the tidal effect changes with the distance between two given bodies but in a somewhat different way.

Let us suppose the Earth and Moon were slowly approaching each other. The total gravitational pull would increase as they moved closer, varying inversely as the square of the distance. If the Moon and Earth were at half their present distance, the gravitational pull between them would be increased 2 × 2, or 4 times. If the Moon and Earth were at one third their present distance, the gravitational pull between them would be increased 3 × 3, or 9 times, and so on.

The tidal effect increases as the total gravitational pull does. The tidal effect increases, *in addition*, for another reason.

The tidal effect depends on the size of the body that is subject to a gravitational field. The larger the size of the body, the greater the tidal effect. However, what counts is not just the size of the body but the size of the body compared with the total distance of the body from the center of gravitational pull.

At the present moment the Moon's diameter of 3,475 is just about 0.009 times the distance between Moon and Earth. If the distance between the two planets were cut in half, the Moon's diameter (which would be the same) would be 0.018 times the distance. In other words as the distance decreased, the tidal effect would increase in proportion to that decrease because the Moon's diameter would make up a larger and larger fraction of the total distance.

You have two factors tending to increase the tidal effects, then—one varying inversely as the square of the distance and the other varying inversely as the distance. If you halve the distance between the Earth and the Moon, the tidal effect would increase 2 × 2 times because of the first factor and 2 times because of the second. The total increase would be 2 × 2 × 2, or 8. Now 2 × 2 × 2 is the cube of 2, so what we are saying is that the tidal effect varies inversely as the *cube* of the distance.

If the distance between two bodies is increased to 3 times what it was, then the tidal effect is decreased to 1/3 × 1/3 × 1/3, or 1/27 what it was. Conversely, when the distance between two bodies is decreased to 1/3 what it was, the tidal effect is increased to 3 × 3 × 3, or 27 times what it was.

If the Earth and Moon, then, were approaching each other, the tidal effect of each on the other would increase steadily and very rapidly. (Whatever the distance, however, the Earth's tidal effect on the Moon would remain 70 times that of the Moon's tidal effect on the Earth.)

Eventually a point would be reached, well before contact was made, where the stretching effect on the Moon would be so enormous that the very structure of the Moon would crack and break. At that moment the Earth, undergoing only 1/70 the tidal effect the Moon was undergoing, would still be able to maintain its integrity, although the enormous ocean tides would undoubtedly destroy everything on the land surface.

In 1849 the French mathematician Edouard A. Roche (1820-1883) showed that if a satellite is held together only by gravitational pull—if it is a liquid for instance—it will break up if it approaches the planet it circles by a distance less than that of 2.44 times the radius of the planet. This is called the *Roche limit*. If a satellite is held together by electromagnetic forces, as our Moon is for instance, it can come a little closer than 2.44 times the radius of the Earth before the tidal stretching overwhelms and destroys it.

The radius of the Earth at the equator is 6,378.5 kilometers, so Earth's Roche limit is about 15,500 kilometers. This is only about 1/25 the actual distance of the Moon. If the Moon were ever to get that close to the Earth, it would break up, and its particles would spread out in orbit around the Earth. It would become a set of rings, like those of Saturn but more massive, and it would no longer exert any important tidal effect on Earth because the various parts of the ring would pull equally in all directions.

The breakup would not continue indefinitely. As the Moon disintegrated into smaller fragments, each fragment, being smaller in size, would experience a smaller tidal effect. Eventually each fragment would be too small for the decreasing tidal effect to break it up further.

If an object is small enough, the tidal effect is insufficient to break it up, even when it is in contact with the attracting body. That is why a spaceship can land on the Moon without breaking up and why we and all the other objects on the Earth's surface can remain intact. The tidal effect for objects of our own size and of the size of the things we work with is insignificant.

The more intense a gravitational field, however, the more intense the tidal effect and the finer the powdering of objects that break up at the Roche limit.

To pass on to gravitational fields more intense than that of the Earth, let us consider the Sun, which is 333,500 times as massive as the Earth and therefore has a gravitational field 333,500 times as intense. The greater diameter of the Sun places its surface farther from its center than Earth's surface is from Earth's center, and since the intensity of the gravitational pull varies inversely as the square of the distance, the surface gravity of the Sun is only 28 times the surface gravity of the Earth.

The tidal effect, however, varies as the inverse cube of distance. Since the Sun's diameter is 109.2 times that of the Earth, we must divide 333,500 (the intensity of the Sun's gravitational field as compared with that of the Earth's) by 109.2 × 109.2 × 109.2, or 1,302,170. Dividing 333,500 by 1,302,170, we get 0.256.

It follows, then, that the tidal effect of the Sun on objects on its surface is only 1/4 the tidal effect of the Earth on objects on its surface.

But suppose the Sun were to contract without losing any mass. Any object on its surface would be closer and closer to its center, and the tidal effect on it would increase rapidly.

Sirius B has a mass equal to the Sun but has a diameter only 1/30 that of the Sun. The tidal effect on the surface of Sirius B would be 30 × 30 × 30, or 27,000 times that on the surface of the Sun, and 7,000 times what it is on Earth's surface.

If we can imagine a human being (two meters tall and weighing 90 kilograms) standing on a white-dwarf star without being affected by its radiation, heat, or total gravity, he would still not be made seriously uncomfortable by its tidal effect, even though that effect is so much larger than on Earth's surface. Multiplying the terrestrial effect by 7,000 would leave the human being stretched by a pull of only about 0.5 kilograms.

What about the Roche limit? Since the Roche limit is 2.44 times the radius of the body exerting the gravitational pull and the cube of 2.44 is 14.53, the tidal effect produced by any body at its Roche limit is 1/14.53 of the tidal effect it produces at its surface. If the tidal effect of Sirius B on its surface is 7,000 times that of Earth at its surface, and if both effects are divided by 14.53, the ratio still stays the same; the tidal effect at Sirius B's Roche limit is 7,000 times that at Earth's Roche limit.

This means that any large object trapped too close to a white dwarf will be broken up much more finely than it will be if it is trapped too close to the Sun or Earth. It also means that small objects that could resist the tidal effects of Sun or Earth at their Roche limits and that would remain whole may nevertheless break up under the influence of a white dwarf.

Let's go further, now, and suppose that an object with the mass of the Sun collapses to the neutron-star stage and is only 14 kilometers in diameter. An object on its surface will now be only 1/100,000 the distance to its center as it would be if it were on the surface of the Sun. The tidal effect on the neutron star's surface is therefore $100,000 \times 100,000 \times 100,000$ times that on the Sun's surface, or a million billion times that on the Sun's surface and a quarter of a million billion times that on the Earth's surface.

A two-meter-tall human being standing on a neutron star and immune to its radiation, heat, or total gravity would nevertheless be stretched apart by a force of 18 billion kilograms in the direction toward and away from the neutron star's center, and of course the human being, or anything else, would fly apart into dust-sized particles. Similarly the neutron star at its Roche limit of 34 kilometers from its center would powder objects finely.

(A second tidal effect arises from the fact that a body on a spherical object has its two sides attracted to the center in slightly different directions. This tends to compress the body from side to side. As long as the body is large enough so that the surface is virtually flat over the width of the body, this effect is very small. Even on a neutron star it is small enough to be ignored—

certainly in comparison with the enormous stretching effect toward and away.)

A human being, even at a distance of 5,000 kilometers from the center of a neutron star would feel a tidal stretch of about 45 kilograms if the long axis of his body were pointing toward the star, and that would be painful indeed.

If a spaceship of the future, effectively shielded against heat and radiation, approached a neutron star at 5,000 kilometers (at which distance it would merely be a dim starlike object to the unaided eye), there would be no need to be concerned about the total gravitational effect. The ship could glide in free fall past the neutron star in a curved orbit and pull away again (if it were moving at a sufficiently high velocity). It would then feel no gravitation, any more than we feel the gravitational pull of the Sun as we, along with the Earth and everything on it, move around the Sun in free fall.

There would, however, be no way of eliminating the tidal effect, and skimming past at 5,000 kilometers would be a harrowing experience. (At closer distances the astronauts would be killed and the ship could break up.)

In 1966 the science-fiction writer Larry Niven wrote an excellent story entitled "Neutron Star," in which the tidal effects of one nearly destroy an unwary astronaut who comes too close. It won a Hugo Award (the science-fiction equivalent of an Oscar) the next year.

Actually, however, the events in the story could not have happened. Tidal effects are no mystery to astronomers and haven't been since the days of Isaac Newton, 300 years ago. Any group of scientists capable of building a spaceship designed to approach a neutron star would certainly understand the danger of the tidal effect, and the astronaut would be certain (barring equipment failure) to remain at a safe distance.

7 BLACK HOLES

FINAL VICTORY

EVEN NOW we are not through.

The nuclear force that keeps neutronium in being can withstand a gravitational inpull intense enough to collapse ordinary atoms and even the electronic fluid. Neutronium can withstand the weight of masses beyond Chandrasekhar's limit. Yet surely, even the nuclear force is not infinitely great. Even neutronium cannot hold up mass endlessly piled on mass.

Since there are stars up to 50 to 70 times as massive as the Sun, it is not inconceivable that once collapse begins, it may on occasion be powered by a gravitational fury even greater and more intense than that which can be withstood by a neutron star. What then?

In 1939, when Oppenheimer was working out the theoretical implications of the neutron star, he took this possibility into consideration, too. It seemed to him that a collapsing star, if massive enough, can contract with such force that even the neutrons will cave in under the impact; even the nuclear force will have to bow to gravitation.

What, then, is the next stopping point of the collapse?

Oppenheimer saw that there is none—no further stopping point. When the nuclear force fails, there is nothing left to withstand gravitation, that weakest of all forces, which when added to and added to by the endless piling together of mass finally becomes the strongest. If a collapsing star crashes through the neutronium barrier, gravitation wins its final victory. The star will thereafter keep on collapsing indefinitely, with its volume shrinking down to zero and its surface gravity increasing without limit.

It appears that the crucial turning point is 3.2 times the Sun's mass. Just as no white dwarf can be more than 1.4 times the Sun's mass without collapsing further, so no neutron star can be more than 3.2 times the Sun's mass without collapsing further.

Any contracting mass that is more than 3.2 times the Sun's mass cannot stop its contraction at either the white-dwarf stage or the neutron-star stage but must go beyond. Furthermore, it appears that any star on the main sequence that is more than 20 times the mass of the Sun will not be able to get rid of enough mass by supernova explosion to make either a white dwarf or a neutron star possible, but must eventually contract to zero. For any star of spectral class O, then, the final victory of gravitation seems a sure thing once the nuclear fuel supply runs out.

(While masses greater than 3.2 times the Sun's mass *must* undergo this ultimate collapse once the process starts, masses less than that *may* do so, as we shall see.)

What happens when this final victory of gravitation takes place and even neutronium gives way? What happens if a neutron star contracts even further?

For one thing the surface gravity of a contracting neutron star goes up steadily, and so does the escape velocity, as the surface of the shrinking object gets nearer and nearer that central point toward which all contraction tends. Already we saw earlier in the book that a neutron star with the mass of our Sun has an escape velocity of 200,000 kilometers per second, which is two-thirds the speed of light.

If the matter in a neutron star continues to contract and the surface gravity grows even more intense, surely there will come a stage where the escape velocity becomes equal to the speed of light. The value of the radius of a body where this is true is called the *Schwarzschild radius* because it was first calculated by the German astronomer Karl Schwarzschild (1873-1916). The zero point at the center is called the *Schwarzschild singularity*.

For a mass equal to that of the Sun, the Schwarzschild radius is just under 3 kilometers. The diameter is equal to twice that, or 6 kilometers.

Imagine, then, a neutron star with the mass of the Sun contracting through the neutron barrier and shrinking from its diameter of 14 kilometers down to one of 6 kilometers. Its density increases thirteenfold and becomes 17,800,000,-

000,000,000 g/cm^3. Its surface gravity is 1,500,-
000,000,000 times that of the Earth, so an average human being would weigh 100 trillion
kilograms if he were standing on such an object.
The tidal effect of such an object is 13 times as
intense as that of a neutron star.

The most important property of such a supercollapsed object, however, is just the fact that
the escape velocity is equal to the speed of light.
(Naturally, if the object collapses to a size still
smaller than the Schwarzschild radius, the escape
velocity becomes greater than the speed of light.)

It so happens that physicists are quite certain
that no physical object possessing mass can move
at a speed equal to or greater than that of light.
That means that any body at the Schwarzschild
radius or less cannot lose any mass by ejection.
Nothing that possesses mass can escape its final
clutch, not even such objects as electrons, which
can, with difficulty, escape from the neutron star.

Things can fall into such a supercollapsed object, but they cannot be ejected again. It is as
though the object was an infinitely deep hole in
space.

What's more, even light or any similar radiation cannot escape. Light consists of massless
particles, so you might think the gravitational
pull of any object, however great that pull might
be, would have no effect on light. By Einstein's
theory of general relativity, however, we know
that light rising against gravity loses some of
its energy and undergoes the Einstein red shift.
This has been an established fact ever since
Adams detected it in connection with Sirius B.
When a collapsed object is at the Schwarzschild
radius or less, light rising from it loses all its

energy and experiences an infinite red shift. This means that nothing emerges.

This supercollapsed object acts not only like a hole but like a black one, since it can emit no light or lightlike radiation. It is in fact called a *black hole* for that reason.

The phrase scarcely seems to be appropriate for an astronomical object whose existence is worked out by abstruse theoretical reasoning. It is too common and everyday a phrase. Another suggested name, therefore, is *collapsar*, a shortened version of *collapsed star*. The dramatic picture of a "black hole" and the very simplicity of its name, however, seems to insure that it will continue to be used.

We have then four types of possibly stable objects:

1) *Planetary objects*, ranging from individual subatomic particles up to masses equal to, say, 50 times that of Jupiter but no more than that. These are all made up (except for individual subatomic particles) of intact atoms, and they generally have overall densities of less than 10 g/cm^3.

2) *Black dwarfs*, which are white dwarfs that have lost so much of their energy that they can no longer shine visibly. These have masses ranging up to 1.4 times the mass of our Sun but no more than that. They are made up of electronic fluid within which are freely moving nuclei and they have densities in the range of 20,000 g/cm^3.

3) *Black neutron stars*, which are neutron stars that have lost so much of their energy that they can no longer shine visibly. These have masses ranging up to 3.2 times the mass of our Sun but no more than that. They are made up

of neutronium, with densities in the range of
1,500,000,000,000,000 g/cm^3.

4) *Black holes*, which yield no light, which
can have masses up to any value, and which are
made up of matter in a state we cannot describe
and with densities of any value up to the infinite.

But are these four varieties of objects truly
stable in the sense that they will undergo no
further change no matter how long a time they
exist?

If a member of any of these four classes were
alone in the universe, as far as we can tell, it
would prove stable and would never undergo
any substantial change. The trouble is, though,
that none of these things are alone in the universe. The universe is a vast mélange of objects
in the different classes of stability, together with
unstable objects such as stars that are evolving
toward one of the latter three classes or, having
reached one of these classes, are still radiating
light en route to final blackness and stability.

What happens then?

Consider the Earth, for instance. It tends to
lose some of its mass as its atmosphere leaks
very slowly away. It also tends to gain some mass
as it collides with and retains meteoric matter
to the tune of some 35,000,000 kilograms a day.
This isn't much, compared with the total mass
of the Earth, but it is considerably greater than
the amount of mass lost by the Earth each day.
We may say, therefore, that the Earth is very
slowly, but steadily, growing more massive.

In the same way the Sun is constantly losing
mass, partly by the conversion of hydrogen to
helium and partly by the ejection of protons and
other particles in the form of a solar wind.

However, it, too, must be gathering dust and meteoric matter from the space it travels through.

The ability to lose mass is true of all objects except black holes. (It is true of black holes, too, under special cases, according to some speculative suggestions, as we shall see.) Even neutron stars eject electrons, or we wouldn't be able to get those microwave pulses. And supernovas eject masses of matter that can be several times the mass of the Sun.

Nevertheless, it can easily be argued that the general tendency in the universe is for large objects to grow at the expense of small. We might imagine, therefore, (simply as an abstract conception) that a planetary object might eventually gain so much mass that it will undergo nuclear ignition and become a star—a very small one, of course—that will eventually reach the white-dwarf stage and finally become a black dwarf.

We might also imagine that after a star has settled down, one way or another, into the presumably stable black-dwarf stage, it might pick up enough mass on its voyage through space to break down the electronic fluid and collapse further to a neutron star. A neutron star, in the same way, might gain enough mass to break down the neutronium and collapse further to a black hole—which, it might seem at first blush, can never lose mass and can only gain mass, with no upper limit to that gain.

There is only one object, then, that would truly appear to be stable through eternity, and that is the black hole. In the end, then—in the long distant end—and always assuming that things will continue to move in the direction in which they now seem to be moving, we might decide that the universe will consist of black holes only —and finally, perhaps, into one black hole containing everything. The entire universe will have collapsed (as I imply in the title of this book).

Or perhaps it isn't quite that simple. We'll get back to the consideration of what the ultimate fate of the universe might be in terms of black holes once we consider their properties somewhat further.

And certainly the first property we ought to consider is the matter of existence. In theory, black holes ought to exist; but in fact, do they exist?

DETECTING THE BLACK HOLE

Detecting a black hole is not easy. White dwarfs, because of their small size and dimness, were far harder to detect, as such, than ordinary stars were. Neutron stars, smaller and dimmer still, were even harder to detect and, if one had to rely on light radiation alone, might never have been detected. It was microwave pulses that gave them away. Obviously a black hole, which emits neither light nor microwaves nor any similar radiation, might baffle observation altogether.

Yet the condition is not entirely hopeless. There is the gravitational field. Whatever happens to the mass that seems to be endlessly added to, and compressed within, a black hole, that mass must remain in existence (as far as we know), and it must continue to be the source of a gravitational field.

To be sure, the total gravitational pull exerted

by a black hole at a great distance is no greater than the total gravitational pull exerted by that mass in any other form. Thus, if you are 100 light-years away from a giant star with 50 times the mass of the Sun, its gravitational pull is so diluted by distance that it is undetectably small. If, somehow, that star becomes a black hole with a mass 50 times the mass of the Sun, its gravitational pull at a distance of 100 light-years will be precisely the same as before and will still be undetectable.

The difference is this: An object can get much closer to a black hole's center than to a giant star's center, so it can experience an enormously more concentrated gravitational pull in the immediate neighborhood of a black hole than it ever can near the far-from-the-center surface of a bloated star of the same mass.

Can the existence of such enormously concentrated gravitational intensities be detected somehow at great distances?

By Einstein's theory of general relativity gravitational activity releases *gravitational waves*, which, in their particle aspect, are spoken of as *gravitons* (just as the particle aspects of light waves are spoken of as photons). Gravitons are far less energetic than photons, however, and cannot conceivably be detectable unless present in unusually high energies, and then just barely. Nothing we know of is likely to produce detectable gravitons—except possibly a large black hole in the process of formation and growth.

In the late 1960s the American physicist Joseph Weber (1919-) used large aluminum cylinders, weighing several tons each and located hundreds of miles apart, as graviton detectors.

Such cylinders would be very slightly compressed and expanded as gravitational waves passed. Weber detected gravitational waves in this manner, and this produced considerable excitement. The easiest conclusion, if Weber's data was correct, was that enormously energetic events are taking place at the center of the Galaxy. A large black hole might be located there.

Other scientists, however, have tried to repeat Weber's findings and have failed, so at this time the question of whether gravitons have been detected or not remains in limbo. There may be a black hole at the center of the Galaxy, but Weber's route to its detection is discounted now, and other ways of detecting black holes must be considered.

One other way, still using the black hole's intense gravitational field in its neighborhood, is to study the behavior of light that might be skimming past a black hole. Light will curve slightly in the direction of a gravitational source; and it will do so detectably, even when it skims past an object like the Sun, with an ordinary gravitational field.

Suppose, now, that there is a black hole lying precisely between a distant galaxy and Earth. The light of the galaxy will pass the pointlike black hole, itself invisible, on all sides. On all sides the light is bent toward the black hole and is made to converge in our direction. This does to light gravitationally what a lens does more conventionally. The effect is, therefore, spoken of as a *gravitational lens*.

If we see a galaxy that despite its distance looks abnormally large, we might suspect it is being magnified by a gravitational lens and that between it and us lies a black hole.

No such phenomenon, however, has yet been observed.

Black holes, however, are not alone in the universe. It could be that there is ordinary matter in the vicinity. If so, sizable objects that happen to approach too closely will be fragmented into dust and, together with matter already in the form of dust and gas, will be circling in an orbit around the black hole as an *accretion disk* about 200 kilometers outside the Schwarzschild radius.

Dust and gas moving in an orbit around the black hole might well stay in such an orbit forever if the individual particles were not interfered with. But mutual collisions bring about a transfer of energy, and some particles, losing energy, spiral inward closer to the black hole and eventually may pass within the Schwarzschild radius, never to emerge again.

On the whole there would be a steady, small inward leak. Inward-spiraling particles, however, lose gravitational energy, which is converted into heat, and they are further heated by the stretching and compression of tidal effects. The result is that they are heated to enormous temperatures and radiate X rays.

Thus, while we cannot detect a bare black hole surrounded by utter vacuum, we might conceivably detect one that is swallowing matter, since that matter will, as its death cry, emit X rays.

The X radiation has to be intense enough to detect across many light-years of space, so it has to represent more than a thin drizzle of occasional dust. There has to be torrents of matter swirling inward, and this means that the black hole has to be in pretty specialized surroundings.

Black holes, for instance, are perhaps most likely to be found where there are huge concentrations of stars in close proximity to one another and where mass buildup might most easily reach the pitch where black-hole formation, sooner or later, will be inevitable.

There are, for instance, globular clusters of stars in which some tens of thousands or even hundreds of thousands of stars are clustered together in a well-packed sphere. Here in our own neighborhood of the universe stars are separated by an average distance of about 5 light-years. At the center of a globular cluster they may be separated by an average distance of 1/2 light-year. A given volume of space in a globular cluster might include 1,000 times as many stars as that same volume in our own neighborhood.

As a matter of fact, a number of globular clusters have been tabbed as X-ray sources, and the possibility is that there are indeed black holes at the center. Some astronomers speculate that such globular-cluster holes may have masses 10 to 100 times that of the Sun.

The central region of galaxies resemble giant globular clusters containing tens of millions or even hundreds of millions of stars. The average separation in the central regions may be 1/10 light-year, and may even diminish to 1/40 light-year at the very center. A given volume of space in a galactic core may have hundreds of thousands, even millions, of stars for every one star present in such a volume in our own neighborhood.

Such crowding doesn't mean stars are bumping one another. Even 1/40 light-year is 40 times the distance between the Sun and Pluto. Still, the chance of violent events would surely increase as the star density in space increases. In recent years there has been increasing evidence of explosions at the centers of galaxies, such energetic explosions that astronomers are at a loss to account for the energies released. Could black holes in some form or other be responsible? Possibly!

Even our own galaxy is not immune. A very compact and energetic microwave source has been detected at the center of our galaxy, and it is tempting to suppose that a black hole is present there. Some astronomers even go as far as to speculate that our galactic black hole has the mass of 100 million stars, so that it must have the mass of 1/1,000 of the entire Galaxy. It has a diameter of 700,000,000 kilometers, which makes it the size of a large red-giant star, but it is something so much more massive that it will disrupt whole stars, tidally, if they venture too close, or gulp them down whole before they can break up, if their approach is rapid enough.

Perhaps every globular cluster and galaxy has a black hole at the center, taking in and never giving out, relentlessly gnawing at normal matter and always growing. Will they swallow up everything eventually? Theoretically, yes, but the rate of doing so may be very small. The universe is 15 billion years old, and yet globular clusters and galaxies still exist unswallowed. There is even a suggestion that central black holes are more nearly the creators of clusters and galaxies rather than their devourers. The black hole may have come first and then served as a "seed," gathering stars about itself as superaccretion disks that become clusters and galaxies.

Constructive as a black hole might have been at first, it is nevertheless swallowing matter now, and however slow the rate, it would not be comfortable to be near one. If indeed there is a black hole at the center of every galaxy, the one closest to us is the one at the center of our own galaxy, and that is 30,000 light-years away. This is a comfortable distance, even with a giant black hole at the other end.

If there is a black hole at the center of every globular cluster, the nearest one to us is in the cluster known as Omega Centauri, which is 22,000 light-years away—still a comfortable distance.

So far, however, black holes at the centers of clusters and galaxies are speculative only. We can't see into a cluster or a galactic core to study its center directly. The vast numbers of peripheral stars hide it, and any indirect evidence we get in the form of X rays or even gravitational waves is not likely to be conclusive in the forseeable future.

Anything else, then?

Suppose we consider not vast conglomerations of stars but merely pairs of them. Suppose we consider binaries.

We can tell the total mass of a binary if its distance from us and the period of its revolution can be determined. If one star looks very small and yet has a large mass, we can tell it is in one stage of collapse or another. That is how the companion of Sirius was detected and how it was finally recognized as a white dwarf.

Suppose, then, we have a binary system in

which both members have collapsed into black holes. The masses, however invisible as a matter of direct observation, are still circling each other and are still, most likely—if young enough—picking up debris from the matter blown off during a supernova explosion. Therefore one would detect a double X-ray in revolution around a center of gravity. Eight X-ray binaries are now known, but as yet the nature of the source in those cases remains unknown.

What if only one star of a binary collapses into a black hole? The companion of that black hole, which could easily be many billions of kilometers distant, will be buffeted by the energy and will find itself circling through a volume of space that is now much dustier than it had been, thanks to the matter ejected in the supernova that preceded the formation of the black hole. The companion may grow warmer as it collects some of this matter and shorter lived in consequence, but for the time being it remains on the main sequence. The gravitational pull to which it is subjected does not increase as a result of the new black hole it has as a partner; rather it is likely to decrease due to the loss of mass in the supernova explosion of its partner.

As viewed from Earth, what one would observe would be a normal star of the main sequence, moving in an orbit about a center of gravity at the opposite side of which was merely an intense source of X rays.

Would these X rays indicate the presence of a neutron star or a black hole? There are differences that might be seized upon for identification. The X rays from a neutron star might be in the form of regular pulses matching the microwave pulses. Two such X-ray pulsars have, in fact, been detected, Centaurus X-3 and Hercules X-1. From a black hole the X rays would vary irregularly as matter is swallowed sometimes in copious quantities, sometimes in sparse quantities. In addition, if such a point source of X rays has a mass of more than 3.2 times that of the Sun, it must be a black hole. (If a mass more than 3.2 times the mass of the Sun should somehow prove to be, incontrovertibly, a neutron star, that would upset the entire theory of black holes. So far, such a too-massive neutron star has not been found.)

In the early 1960s, when X-ray sources were first discovered in the sky, a particularly intense source was located by rocket observation in 1965 in the constellation Cygnus. The X-ray source was named Cygnus X-1.

In 1969 an X-ray detecting satellite was launched from the coast of Kenya on the fifth anniversary of Kenyan independence. It was named Uhuru from the Swahili word for "freedom." It multiplied knowledge of X-ray sources to unlooked-for heights, detecting 161 such sources, half of them in our own galaxy and 3 of them in globular clusters.

In 1971 Uhuru detected a marked change in X-ray intensity in Cygnus X-1, which virtually eliminated it as a possible neutron star and raised the possibility of a black hole. Now that attention was eagerly focused on Cygnus X-1, microwaves were also detected, and this made it possible to pinpoint the source very accurately and place it in just next to a visible star.

This star was HD-226868, a large, hot blue star of spectral class B, some 30 times as massive

as our Sun. C. T. Bolt of the University of
Toronto showed HD-226868 to be a binary. It is
clearly circling in an orbit with a period of 5.6
days—an orbit the nature of which makes it
appear that the other star is perhaps 5 to 8
times as massive as the Sun.

The companion star cannot be seen, however,
even though it is a source of intense X rays.
If it cannot be seen, it must be very small. It is
too massive to be either a white dwarf or a neutron star, and the inference seems to be, then,
that the invisible star is a black hole.

Furthermore, HD-226868 seems to be expanding as though it were entering the red-giant
stage. Its matter would therefore be spilling over
into the black-hole companion, which would explain why the latter is so intense an X-ray source.

This is still rather indirect evidence, and not
all astronomers agree that Cygnus X-1 is a black
hole. A lot depends on the distance of the binary.
The greater the distance, the greater the mass
required of the stars in order to have them have
so short an orbital period, and the more likely
Cygnus X-1 is massive enough to be a black
hole. Some astronomers maintain that the binary
is considerably closer than the 10,000 light-years
its distance is usually estimated to be and that
Cygnus X-1 is therefore not a black hole. The
consensus, however, seems (at least so far) to
favor the black-hole hypothesis.

A few other binaries have since been observed
in which one of the pair may be a black hole.
These include X-ray sources known as X Persei
and Circinus X-1.

There are also black-hole possibilities where
X-ray emission isn't a factor. In some cases you

can deduce a very close binary by the behavior of the spectral lines. Epsilon Aurigae, from the behavior of its spectral lines, seems to be revolving around an invisible companion, Epsilon Aurigae B. What's more, the spectroscope data makes it seem that Epsilon Aurigae A, the visible star, has a mass 17 times that of the Sun, while Epsilon Aurigae B, the invisible one, has a mass 8 times that of the Sun. Again the combination of invisibility and great mass indicates the possibility that Epsilon Aurigae B is a black hole (though some astronomers maintain Epsilon Aurigae B is invisible because it is a new star in the process of formation, and has not yet ignited).

MINI-BLACK HOLES

If black holes exist merely at the centers of galaxies, then there would be only one in our galaxy. If they existed also at the center of globular clusters, there would be perhaps 200 of them in our galaxy. However, if they also exist as part of ordinary binary systems, there is the potentiality of vast numbers of them. After all, there are tens of billions of binaries in our galaxy.

What's more, they need not be part of binaries only. It so happens that the nearby companion gives away the existence of a black hole, which is why we think of them in connection with binaries. Black holes might also evolve from single stars, and then, without nearby matter to produce the X rays and a nearby companion to offer a measurement of mass, they might be impossible to detect, but they would be there just the same.

Taking all this into account, some astronomers suspect there may be as many as a billion star-sized black holes in any galaxy like ours. If this is true and if the black holes are more or less evenly distributed, the average distance between them is 40 light-years and any particular star might be, on the average, 20 light-years from some black hole or other.

Of course, it is more likely that the black holes are distributed as unevenly as the stars themselves are. Ninety percent of all the stars in our galaxy (or in any similar galaxy) are located in the relatively small central regions. Only 10 percent are in the voluminous, but sparsely populated, spiral arms, where our own Sun is located. It might be, then, that only 10 percent of the black holes of our galaxy are located in the spiral arms, that they are well spread out here, and that it is likely that the nearest black hole to us is several hundred light-years away.

Of course, in talking about black holes, we have so far been talking about black holes with masses equal to those of massive stars, and there are indeed astronomers who think that the average black hole has a mass 10 times that of our Sun.

It might seem that anything much less couldn't exist, since only star-sized objects could possess a gravitational field large enough to produce a compression intense enough to break through the neutronium barrier and produce a black hole.

According to Einstein's theory of general relativity, however, black holes can come in all sizes. Any object possessing mass, no matter how small that mass may be, also possesses a gravitational field. If the object is compressed into a smaller

and smaller volume, that gravitational field becomes more and more intense in its immediate
vicinity and eventually becomes so intense that
the escape velocity from its surface is greater
than the speed of light. It has, in other words,
shrunk within its Schwarzschild radius.

The Earth would become a black hole if it
shrank to a diameter of 0.87 centimeter (the size
of a large pearl). A mass the size of Mount
Everest would become a black hole if it shrank
to the size of an atomic nucleus.

We might go on in this way until we reach
the smallest mass known, that of an electron,
but there are subtle theoretical reasons for supposing that masses less than 10^{-5} grams may be
unable to form black holes. A mass of 10^{-5} grams
(a speck of matter just visible to the eye) would
become a black hole if it were reduced to a diameter of something like 10^{-33} cm, at which time it
would have a density of 10^{94} g/cm^3. (At such a
density an object the size of an atomic nucleus
would have a mass equal to that of the entire
universe.)

But what can possibly compress small objects
into such *mini-black holes*. It can't be their own
gravitational fields, so it must be some compressing force from outside. But what force from outside can be strong enough to produce them?

In 1971 the English astronomer Stephen
Hawking suggested that one conceivable force
would have come at the time the universe was
formed—the force of the big bang itself. With
vast quantities of matter exploding all over the
place, some different sections of the expanding
substance might collide. Part of this colliding
matter might then be squeezed together under

enormous pressure from all sides. The squeezed matter might shrink to a point where the mounting gravitational intensity would keep it shrunk forever.

There is, of course, no evidence whatsoever that such mini-black holes exist, not even as much evidence as what Cygnus X-1 supplies for star-sized black holes. What's more, some astronomers scout the whole idea and think that there exists only black holes with masses distinctly greater than that of our Sun.

Nevertheless, *if* mini-black holes exist, then it is likely that there are many more of these than of the star-sized ones. Can it be, then, that if there are star-sized black holes spread at average separations of 40 light-years, there might be a whole array of moderate-sized to microscopic-sized black holes at much closer intervals? Might space be littered with them? Hawking thinks there may be as many as 300 per cubic light-year in the universe.

It is important to remember that there is no indication of this whatever. But then, if mini-black holes are thickly spread in space, the total gravitational effect is tiny, and it can be detected only in the immediate neighborhood of the object —a few kilometers away, a few centimeters away, a few micrometers away, depending on its size.

To be sure, such tiny black holes must be ceaselessly growing, for they will engulf any dust particle with which they might collide—at least that is the usual view of matters. (Hawking also advances subtle reasons for supposing that mini-black holes can lose mass, and that really small ones might "evaporate" and explode before they can gain much mass.)

If a mini-black hole collides with a larger body, it will simply bore its way through. It will engulf the first bit of matter with which it collides, liberating enough energy in the process to melt and vaporize the matter immediately ahead. It will then pass through the hot vapor, absorbing it as it goes and adding to the heat, emerging at last as a considerably larger black hole than it was when it entered.

(If a mini-black hole enters a larger body that has very little in the way of energy of motion, it may become trapped within the body and sink, eventually, to its center where it may gradually eat a hole out for itself and continue to grow at an ever slower rate like a parasite consuming its host.)

To be sure, the volume of such mini-black holes is so tiny, the total gravitation so small, and the volume and emptiness of space so enormous that collisions must be rare indeed. In the whole 15 billion years since the big bang the vast majority of the tiny black holes must have gained so little mass that they are still tiny black holes and still just about impossible to detect.

In the face of the olds, of course, a mini-black hole might collide with the Earth. The heat produced as it passes through the atmosphere would be enough to produce spectacular effects that people couldn't help noticing, and its passage through the Earth might produce effects as well.

Has it ever happened?

We don't know. There are no signs that we know of that anything like this happened in prehistoric times, but can we be sure? Was Sodom destroyed because a mini-black hole struck? How can we tell? The destruction might have been caused by an ordinary meteorite, a volcanic eruption, and earthquake, or the whole tale may be mythical. The records aren't good enough.

Has anything happened in historic times that might be accounted for by a mini-black hole? One thing!

On June 30, 1908, what was at first thought to be a large meteor strike occurred in the Tunguska region of central Siberia. Every tree for 30 kilometers in every direction was knocked down, and an entire herd of 500 reindeer was destroyed. In later years thorough searches of the area found no craters and no meteor fragments.

Researchers decided that the explosion must have taken place in the atmosphere. Some thought it might have been a small comet made up of icy materials that melted and vaporized in the passage through the atmosphere, creating a huge bang and peppering the Earth with fragments of gravel (embedded in the ice) in such a way that no noticeable gouges appeared.

Others thought it might have been an example of antimatter that struck the Earth. Antimatter is made of material resembling ordinary matter except that all the subatomic particles composing it are opposite in properties to the particles composing ordinary matter. Antimatter interacts with matter, converting everything on both sides into energy. A particle of antimatter striking the normal matter of Earth will disappear, taking an equal mass of normal matter with it and producing a bang equal to that of a hydrogen bomb with a nuclear warhead some 15 or more times as massive as itself.

It has even been suggested that the bang was caused by the wreck of a nuclear-powered spaceship manned by extraterrestrial astronauts.

One other suggestion, however, was that it was a mini-black hole that did it, one that struck, created a vast explosion as it passed through the atmosphere, entered the Earth at an angle, passed through and absorbed more matter, and emerged at last in the North Atlantic Ocean, where it produced a gigantic water spout and explosion that went unseen and unheard by man. It then proceeded back into space, considerably larger than when it arrived but still a mini-black hole.

Of course, this mini-black-hole suggestion is just speculation, too. Some astronomers point out that a mini-black hole passing through the body of the Earth and out the ocean might well have set off earthquakes and should surely have initiated a tidal wave—yet neither event took place in conjunction with the strike of 1908.

There is simply no way, as yet, of either proving or disproving the mini-black-hole explanation of the 1908 event. There may never be a way unless a similar event happens again at the time when scientists, with their knowledge of the universe vastly advanced over what it was in 1908, can study the event as it occurs.

THE USE OF BLACK HOLES

Naturally, any scientist, however dedicated, cannot view the possibility of a collision between a mini-black hole and the Earth with satisfaction. If the 1908 event had not fortunately struck one of the few areas of Earth's land surface on which no human beings lived for many kilometers in every direction, there might have been fearful human and property destruction.

One can easily imagine such a strike utterly wiping out Washington, D.C., or Moscow, for instance, if it happened to be unfortunately aimed. The results might so resemble the strike of a hydrogen bomb that whichever superpower was struck might launch a retaliatory strike before learning the truth, and the whole Earth might be ravished.

Of course, I can't repeat often enough that the Siberian strike might not have been caused by a mini-black hole; that there may be no such things as mini-black holes; that if there are, the chances of collision may be far less than that of being struck by a meteorite while you are asleep in your bed.

Still—what if mini-black holes exist?

We might eventually learn to protect ourselves against them. If human beings ever reach the stage where they have observatories and colonies on other worlds of the solar system and in artificial structures in space itself, there may come an opportunity to study mini-black holes in their native haunt, so to speak, under conditions that don't involve a collision with Earth.

In fact, we can even dream that techniques will be developed to capture a black hole by means of its gravitational field (very intense in its immediate neighborhood but quite small in total) and force it to veer in its flight by just enough to effect a miss if it was otherwise heading for Earth. That would be a side effect of space exploration that might well be worth any amount of money spent on it.

Those who speculate far in advance of the present capabilities of science and who enjoy dreaming up fantastic visions of the future * might even find it possible to hope that we are relatively close to a black hole (though far enough away to be safe).

A black hole is, after all, a gateway to enormous energies. Any object spiraling into it will in the process radiate a great deal of energy.

Most of the energy in any object resides in its mass, since each gram of mass is the equivalent of 9×10^{20} ergs of energy. The energy we get by burning oil or coal, for instance, makes use of only a tiny fraction of 1 percent of the mass of the fuel. Even nuclear reactions liberate only a couple of percent of the mass. An object spiraling into a black hole or, under certain conditions, skimming it without actually entering it may convert up to 30 percent of its mass into energy.

What's more, only certain substances can be burned to yield energy; only certain atomic nuclei can be split or fused to yield energy. Anything, however—*anything*—will yield energy on falling into a black hole. The black hole is a universal furnace, and everything that exists and has mass is its fuel.

Perhaps we can imagine some far-advanced civilization of the future tapping the black hole for the energy it can produce, stoking it with asteroids as we might stoke an ordinary furnace with coal. In that case, if the galaxy possesses hundreds, or even thousands, of advanced civil-

---

* This includes myself, since (as the reader may know) I am a science-fiction writer of some repute.

izations (as some astronomers suspect it might), it may be those that are reasonably close to sizable black holes who may have the richest supply of available energy and who flourish as earthly nations do when they are rich in energy resources.

To be sure, it is exceedingly unlikely we will find massive black holes that we can use as a universal furnace. Nor might we really be anxious to find one within too few light-years of ourselves, since the larger they are, the more unmanageable they are.

Perhaps it is better, until such time as our technology advances sufficiently, to make do with one of the much more common (if they exist at all) mini-black holes and make use of more conventional means of gaining energy.

Suppose we find a mini-black hole somewhere in the solar system passing through or, even better, orbiting the Sun. We might in each case seize it by its gravitational field, tug it in the wake of some massive object, and set it up in an orbit around the Earth (if a nervous humanity will allow it).

A stream of frozen hydrogen pellets can then be aimed past the mini-black hole so that it skims the Schwarzschild radius without entering it. Tidal effects will heat the hydrogen to the point of fusion, so that helium will come through at the other end. The mini-black hole will then prove the simplest and most foolproof nuclear-fusion reactor possible, and the energy it produces can be stored and sent down to Earth.

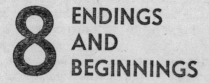

8 ENDINGS AND BEGINNINGS

THE END?

WE ARE BOUND to be curious about what can possibly happen to matter that falls into a black hole.

It is very difficult to satisfy that curiosity. Indeed, all we can do is speculate, for we have no way of telling whether any of the laws of nature that have so painstakingly been worked out by observing the universe around us can hold under the extreme conditions of a black hole. We can't duplicate those conditions in any way here on Earth, and we can't observe those conditions in the heavens, since we know of no black hole in our vicinity.

It follows, then, that we can only assume that the laws of nature do hold and then try to speculate what might happen.

One thing that might happen is that the worst

does *not* happen or at least is not observed to happen. How, for instance, can mass compress down to zero volume and infinite density at the Schwarzschild singularity? This so boggles the mind that we must search for something that will prevent it.

For instance, Einstein's theory makes it seem that increasing intensity of gravity has the effect of slowing the passage of time. This is not something we can observe in the universe very easily, for outside of black holes and neutron stars, those gravitational intensities we encounter have only a negligible effect on the time rate.

Because of this, if we could observe something dropping down into a black hole, we would see it move more and more slowly as it approaches the Schwarzschild radius, creeping ever more slowly until at the Schwarzschild radius we would see it stop dead. However, as it approaches, the Einstein red-shift, also dependent on gravitational intensity, robs light and lightlike radiation of more and more of its energy. The object dropping downward will grow dimmer as it moves more slowly, and at the Schwarzschild radius, where it freezes, it also blacks out. The result is that we cannot possibly observe anything within the Schwarzschild radius.

If we imagine an astronaut falling into a black hole and somehow retaining consciousness and the ability to be aware of his surroundings, he would feel no change in time rate; that change is only something an outsider would see as existing.

The astronaut falling into a black hole would pass through the Schwarzschild radius without knowing it was any kind of barrier, and he would keep on falling toward the singularity ahead. However, one way of interpreting the events that follow is to suppose that from the standpoint of the astronaut the distance before him would expand as he fell, so that though he might fall forever, he would never reach the center. The black hole in that view is a bottomless hole.

Although in either way of looking at objects falling into a black hole, there is no reaching the center, no zero volume, no infinite density—yet there is also no turning back. The fall is irreversible, so once again let us consider the possible end of the universe.

If there is truly no way to reverse or neutralize the black hole, then those that exist now can only grow; and new ones may form.

If there is a black hole at the center of every galaxy and at the center of every globular cluster, then in the end (however long delayed) each galaxy will become a large black hole surrounded by satellite black holes that are much smaller.

Two black holes can collide and coalesce, but a black hole once formed cannot split up. Therefore we might imagine that sooner or later the globular cluster black holes in their orbit around the galactic black hole may coalesce with one another and eventually with the central one, so that, given enough time, the galaxy will be one black hole only.

Galactic units may consist of one galaxy only, but they may also consist of several galaxies (in extreme cases, several thousand) that are bound together by gravitational attraction. Each galaxy in a unit may be a black hole, and these may coalesce, too.

Can we go on to suppose that all the black

holes in the universe will eventually coalesce into one universal black hole?

Not necessarily. The universe is expanding, so the galactic units (either single galaxies or galactic clusters) are steadily increasing their distance from one another. Most astronomers seem to feel that this will continue indefinitely into the future. If so, we have the vision of a universe consisting of billions of black holes, each with a mass of anywhere from millions to trillions times that of our Sun, moving endlessly away from one another.

The very act of expanding, however, may just possibly introduce a change.

Back in 1937 the English physicist Paul Adrien Maurice Dirac (1902-) advanced the startling suggestion that the intensity of the gravitational force generally depends upon the overall properties of the universe. The greater the average density of the universe, the stronger the gravitational force is, relative to the other forces of the universe.

Since the universe is expanding, the average density of matter is decreasing as it spreads out over a steadily greater volume. It is because of the great expansion that has taken place so far (in this view) that the gravitational force is so weak in comparison with the others, and as the universe continues to expand, it will grow weaker still.

Dirac's suggestion has not yet been observed to be true, and many physicists suspect that the gravitational constant (the value of which dictates the basic strength of the gravitational force) is not only the same everywhere in space but does not vary with time, either. Nevertheless, if Dirac's suggestion should prove to be true, it alters the picture just described.

As the universe expands and gravitation grows ever weaker, those objects held together primarily by gravitational force will expand and become less compact and dense. This will include white dwarfs and neutron stars that have already formed, and it will also include black holes. The tendency will be for all objects to bloat into matter held together by the electromagnetic force or not held together at all. Even black holes will, little by little, disgorge, and in the end the universe will be a vast, incredibly thin cloud of gravel, dust, and gas growing endlessly vaster and thinner.

If this is so, it might seem that the universe began as a huge mass of compressed matter and will end as a huge volume of thin matter.

This raises the puzzle of where the compressed matter came from. We needn't worry about the matter as such, for it is just a very compact form of energy, and we might suppose that the energy has always existed and always will exist —much of it in the form of matter. The question is, how did the matter come to be compressed into the cosmic egg to begin with?

We might suppose that if we consider the universe to progress from compressed to expanded, we are taking into account only half the life cycle.

Suppose the universe began as an endlessly thin volume of gravel, dust, and gas. Slowly, over incredible eons, it condensed until it formed the cosmic egg, which then exploded and over equally incredible eons it has been restoring matter as it had been. We happen to be living during the period shortly after (a mere 15 billion years after) the explosion.

Yet somehow the thought of the universe as a one-shot seems vaguely unsatisfactory. If dispersed matter could collect itself, coalesce, contract, and finally form a cosmic egg, then why may not the dispersed matter that forms as the end product of the cosmic-egg explosion (whether it consists of black holes or of dispersed matter) collect itself again, contract once more, and form a second cosmic egg?

Why may not this be repeated over and over? Why, in short, might there not be an endlessly *oscillating universe*?

Astronomers have worked out those conditions that are required to produce an oscillating one. The choice depends on something like escape velocity. There is a certain gravitational force among the galactic units of the universe generally, and there is an escape velocity associated with that force. If the universe is expanding outward at a velocity greater than the escape velocity, then it will expand forever and will never contract. If it is expanding at less than the escape velocity, then the present expansion must eventually come to a halt, and the contraction must then begin.

But is the present observed velocity of expansion larger or smaller than the escape velocity? That depends on the value of the escape velocity, which depends on the value of the overall gravitational force among the galactic units, which depends, it turns out, on the average density of matter in the universe.

The greater the average density of matter in the universe, the greater the gravitational force

among the galaxies, the greater the escape velocity—and the greater the likelihood that the
present velocity of expansion is not greater than
the escape velocity and that the universe will
oscillate, that it is *closed*.

Naturally it is difficult to determine the average
density of the universe, since it is hard to determine how much total mass is present in a
large enough volume of it to be representative of
the whole. Making use of the best data available,
some astronomers seem quite convinced that the
average density is only about 1/100 the value
needed for oscillation, that the universe is *open*
and is doomed to expand forever. (If the gravitational force is weakening as the universe expands, then an even greater average density is
required for oscillation, and the apparent density
falls even further short of that requirement.)

And yet although the arguments against a
closed and oscillating universe seem strong, can
they really be the last word? Clusters of galaxies
that seem to be held together by gravitational
pull nevertheless don't seem to have sufficient
mass to supply that pull. They should be flying
apart in response to the general expansion of the
universe, and yet they do not seem to be doing
so. There is thus what is called the problem of
the missing mass.

Can that missing mass consist of black holes?
Except in a very few cases there is no way of
detecting black holes, and we don't have the fog-
giest notion how much mass is tied up inde-
tectably in those black holes of all sizes. It seems
difficult to believe that black holes account for a
hundred times as much mass as do all the visible
objects of the universe. Yet we are on the very

borderline of what we can observe and reason out, and we can't afford to be too certain one way or the other. The evidence *seems* to point to an open, over-expanding universe, but it *may be* that, counting the black holes, there is enough mass to keep the universe closed and oscillating after all.

WORMHOLES AND WHITE HOLES

The discomfort over an open, ever-expanding, one-time-only universe is such that astronomers seem to twist and turn in an effort to get away from the evidence that points to it.

Back in 1948 Thomas Gold, along with the English astronomers Fred Hoyle and Hermann Bondi, tried to get around it by suggesting what came to be called the *continuous creation* universe. The thought was that matter would be created continuously, an atom at a time, here and there in the universe. It would be created at a rate so low that we couldn't detect it.

Nevertheless, as the universe expanded and the space between galactic units increased, enough matter would be formed to collect into new galaxies in that space between. On the whole just enough galaxies would be formed to make up for the spreading apart of the old ones. The universe would be a vast mélange of galaxies ranging from those just forming through all the stages of development to those just dying. The universe would be infinitely large in space and eternally enduring in time. Stars and galaxies would be born and would die, but the universe as a whole would be immortal and would neither come into being nor go out of being.

This was an attractive theory, but the evidence in its favor was almost nonexistent and never grew any stronger. In fact it grew weaker. If the continuous creation universe were what actually existed, then there would never have been a big bang. For that reason any evidence that seemed to substantiate the big bang tended to wipe out continuous creation.

In 1964 American physicist Robert Henry Dicke (1916-) pointed out that the big bang, if it took place 15 billion years ago, must have left traces that should even now be visible 15 billion light-years away (for it takes light 15 billion years to get here from that distance, and so the light of the big bang is just arriving now).

The big-bang radiation, of a very energetic and short-wave type, has shifted, because of this vast distance, far toward the low-energy red end of the spectrum. It has shifted past the red and into the much longer, lower-energy microwave section of the spectrum. Since the big bang must be visible 15 billion light-years away in any direction, the microwaves must come from all parts of the sky as a *background radiation*.

In 1965 two scientists at Bell Telephone Laboratories, Arno A. Penzias and Robert W. Wilson, demonstrated the existence of a faint background radiation with just the characteristics Dicke had predicted. The big bang had been detected, and continuous creation has been (at least for now) killed.

That route for avoiding the open universe has failed. There are, however, others, and for those let us return to the black holes.

So far we have been talking about black holes that have only one property—mass. The nature of the mass does not matter. If a kilogram of platinum, or a kilogram of hydrogen, or for that matter a kilogram of living tissue is added to a black hole, what is added is a kilogram of mass without any history of its previous state.

There are two other properties, and only two, that can be possessed by a black hole. One is electric charge, and the other is angular momentum. That means that any black hole can be described completely by measuring its mass, electric charge, and angular momentum. (It is possible for electric charge and angular momentum to be each zero; but mass cannot be zero, or it is not a black hole.)

While a black hole may have an electric charge, it can only have it if the mass that formed the black hole to begin with or that is added to it afterward has an electric charge. In point of fact the electric charges, positive and negative, in sizable pieces of matter tend to be equal in quantity, so the overall charge is zero. Consequently, black holes are quite likely to have essentially zero charge.

Not so with angular momentum. There, indeed, the situation is reversed, and it is quite likely that every black hole has a considerable angular momentum.

Angular momentum is a property of any object rotating on its axis, or revolving around an outside point, or both. Angular momentum includes both the speed of rotation or revolution of the object and the distance of its various parts from the axis or center around which it turns. The total angular momentum of a closed system (one in which no angular momentum can leak in or out) must be conserved; that is, it can neither be increased nor decreased.

This means that if the distance is increased, the speed of turning must be decreased, and vice versa. An ice skater takes advantage of this when he sets up a spin with his arms out-stretched. He draws his arms in toward his body, decreasing the average distance of the parts of his body from the axis of rotation, and his rate of spin increases markedly. He extends his arms again, and he slows down at once.

Every star we know of rotates about its axis and therefore has a large amount of rotational angular momentum. When a star collapses, to make up for that, its speed of rotation must increase. The more extreme the collapse, the greater the gain in speed of rotation. A brand-new neutron star can spin as much as a thousand times a second. Black holes must spin more rapidly still. There's no way of avoiding that.

We can say, then, that every black hole has mass and angular momentum.

The mathematical analysis of Schwarzschild applied only to nonrotating black holes, but in 1963 the astronomer Roy P. Kerr worked out a solution for rotating black holes.

In rotating black holes the Schwarzschild radius is still there, but outside it is a *stationary limit*, which forms a kind of equatorial bulge around the black hole as though it were something pushed outward by the centrifugal effect.

An object falling within the stationary limit but remaining outside the Schwarzschild radius is semitrapped. That is, it can still get out, but only under special circumstances. If it happens to move with the direction of the turn, the rotating black hole will tend to drag the object around like a stone in a sling and to hurl it back out beyond the stationary limit with more energy than it entered with. The additional energy is at the expense of the black hole's rotation. In other words angular momentum is transferred from the black hole to the object, and the black hole slows down.

In theory up to 30 percent of the entire energy of a rotating black hole can be milked out of it by carefully sending objects through the stationary limit and collecting them on the way out, and this is another way in which some advanced civilizations might use black holes as an energy source.* Once all the rotational energy is gone, the black hole has only mass; the stationary limit coincides with the Schwarzschild radius. The black hole is then said to be "dead," since no further energy can be obtained from it directly (though some can be obtained from matter as it spirals into it).

Even stranger than the possibility of stripping rotational energy from the black hole is that the Kerr analysis offers a new kind of end for matter entering a black hole. This new kind of end was foreshadowed by Albert Einstein and a co-worker named Rosen some 30 years earlier.

The matter crowding into a rotating black hole (and it is very likely that there is no other kind) can, in theory, squeeze out again somewhere else, like toothpaste blasting out of a fine hole

---

* Not all astronomers agree with this concept of stripping the rotational energy of a black hole. In fact almost anything some astronomers suggest about a black hole is denied by other astronomers. We are here at the very edge of knowledge, and everything, one way or the other, is very uncertain and iffy.

in a stiff tube that is brought under the slow pressure of a steamroller.**

The transfer of matter can apparently take place over enormous distances—millions or billions of light-years—in a trifling period of time. Such transfers cannot take place in the ordinary way, since in space as we know it the speed of light is the speed limit for any object with mass. To transfer mass for distances of millions or billions of light-years in the ordinary way takes millions or billions of years of time.

One must therefore assume that the transfer goes through tunnels or across bridges that do not, strictly speaking, have the time characteristics of our familiar universe. The passageway is sometimes called an *Einstein-Rosen bridge*, or, more colorfully, a *wormhole*.

If the mass passes through the wormhole and suddenly appears a billion light-years away in ordinary space once more, something must balance that great transfer in distance. Apparently this impossibly rapid passage through space is balanced by a compensating passage through time, so that it appears 1 billion years ago.

Once the matter emerges at the other end of the wormhole, it expands suddenly into ordinary matter again and, in doing so, blazes with radiated energy—the energy that had, so to speak, been trapped in the black hole. What we have emerging, then, is a *white hole*, a concept first suggested in 1964.

If all this is really so, white holes, or at least some of them, might conceivably be detected.

That would depend, of course, upon the size

** This suggestion, too, is denied by some astronomers.

of the white hole and upon its distance from us. Perhaps mini-black holes form mini-white holes at vast distance, and we would surely never see them. Huge black holes would form huge white holes, however, and these we might see. Are there any signs of such white holes?

There may be—

QUASARS

In the 1950s sources of radio waves were detected that on closer inspection seemed to be very compact, emerging from mere pinpoint sections of the sky. Ordinarily, radio sources found in those early days of the science were from dust clouds or from galaxies and were therefore more or less spread out over a portion of the sky.

Among those compact radio sources were several known as 3C48, 3C147, 3C196, 3C273, and 3C286. (Many more have been discovered since.) The *3C* is short for *Third Cambridge Catalog of Radio Stars*, a list compiled by the English astronomer Martin Ryle (1918-).

In 1960 the areas containing these compact radio sources were investigated by the American astronomer Allan Rex Sandage (1926-), and in each case something that looked like a dim star seemed to be the source. There was some indication that they might not be normal stars, however. Several of them seemed to have faint clouds of dust or gas about them, and one of them, 3C273, showed signs of a tiny jet of matter emerging from it. In fact there are two radio sources in connection with 3C273, one from the star and one from the jet.

There was some reluctance, therefore, to call these objects stars, and they were instead described as *quasi-stellar* (starlike) *radio sources.* In 1964 Hong-Yee Chiu shortened that to *quasar,* and that name has been kept ever since.

The spectra of these quasars were obtained in 1960, but they had a pattern of lines that were completely unrecognizable, as though they were made up of substances utterly alien to the universe. In 1963, however, the Dutch-American astronomer Maarten Schmidt (1929-) solved that problem. The lines would have been perfectly normal if they had existed far in the ultraviolet range. Their appearance in the visible-light range meant they had been shifted a great distance toward the longer wavelengths.

The easiest explanation for this was that the quasars are very far away. Since the universe is expanding, galactic units are separating, and all seem to be receding from us. Therefore, all distant objects have their spectral lines shifted toward the longer waves because that is what is to be expected when a source of light is receding from us. Furthermore, since the universe is expanding, the farther an object, the faster it is receding from us and the greater the shift in spectral lines. From the spectral shift, then, the distance of an object can be calculated.

It turned out that the quasars were billions of light-years away. One of them, OQ172, is about 12 billion light-years away, and even the nearest, 3C273, is over a billion light-years away and farther than any nonquasar object we know about. There may be as many as 15 million quasars in the universe.

A quasar is a very dim object, as we see it,

but for it to be visible at all at those enormous distances, it must be exceedingly luminous. The quasar 3C273 is 5 times as luminous as our galaxy, and some quasars may be up to 100 times as luminous as the average galaxy.

Yet, this being so, if quasars were simply galaxies with up to a hundred times as many stars as an average galaxy and therefore that much brighter, they ought to have dimensions large enough to make them appear, even at their vast distances, as tiny patches of light and not as starlike points. Thus, despite their brightness they must be more compact than ordinary galaxies.

As early as 1963 the quasars were found to be variable in the energy they emitted, both in the visible-light region and in the microwave region. Increases and decreases of as much as three magnitudes were recorded over the space of a few years.

For radiation to vary so markedly in so short a time, a body must be small. Such variations must involve the body as a whole, and if that is so, some effect must make itself felt across the full width of the body within the time of variation. Since no effect can travel faster than light, it means that if a quasar varies markedly over a period of a few years, it cannot be more than a light-year or so in diameter and may be considerably smaller.

One quasar, 3C446, can double its brightness in a couple of days, and it must therefore be not more than 0.005 light-year (50 billion kilometers) in diameter, or less than five times the width of Pluto's orbit around the Sun. Compare this with an ordinary galaxy, which may be

100,000 light-years across and in which even the dense central core may be 15,000 light-years across.

This combination of tiny dimensions and enormous luminosity makes the quasars seem like a class of objects entirely different from anything else we know. Their discovery made astronomers aware of the possibility of hitherto unknown large-scale phenomena in the universe and spurred them on, for the first time, to consider such phenomena, including the black hole.

And it is conceivable that there is a link between black holes and quasars. The Soviet astronomer Igor Novikov and the Israeli astronomer Yuval Ne'eman (1925-) have suggested that quasars are giant white holes at the other end of a wormhole from a giant black hole in some other part of the universe.*

But let's take another look at quasars. Are they really unique, as they seem to be, or are they merely extreme examples of something more familiar?

In 1943 a graduate student in astronomy, Carl Seyfert, described a peculiar galaxy, which has since been recognized as one of a group that are now termed *Seyfert galaxies*. They may make up 1 percent of all known galaxies (meaning as many as a billion altogether), though actually only a dozen examples have been discovered.

In most respects Seyfert galaxies seem normal and are not unusually distant from us. The cores of the Seyfert galaxies, however, are very compact, very bright, and seem unusually hot and active—rather quasarlike in fact. They show

---

* This is purely speculative, of course, and the remainder of the book is almost entirely speculation, some of it my own.

variations in radiation that imply the radio-emitting centers at their core are no larger than quasars are thought to be. One Seyfert galaxy, 3C120, has a core that makes up less than one eighth the diameter of the galaxy as a whole but is three times as luminous as all the rest of the galaxy combined.

The strongly active center would be visible at greater distances than the outer layers of the Seyfert galaxy would be, and if such a galaxy were far enough, all we would see by either optical or radio telescopes would be the core. We would then consider it a quasar, and the very distant quasars may simply be the intensely luminous nuclei of very large, very active Seyfert galaxies.

But then consider the core of a Seyfert galaxy —very compact, very hot and active. One Seyfert galaxy, NGC 4151, may have as many as 10 billion stars in a nucleus only 12 light-years across.

These are just the conditions that would encourage the formation of black holes. Perhaps the mere fact that a certain volume of space is subject to black-hole formation may also make it subject to the blossoming out of a white-hole.

We can imagine black holes forming here and there in the universe, each producing an enormous strain in the smooth fabric of space. Wormholes form between them, and matter may leak across at a rate slow in comparison with the total quantity in the black hole serving as source but large enough to produce enormous quantities of radiation in some cases. The rate of matter flow may vary for reasons we do not as yet understand, and this may bring about the variations in the brightness of quasars.

There may be many white holes of all sizes, each connected to its black hole (which itself may come in any size), and we may be aware only of the giant-sized ones. It may be that if all black holes/white holes were taken into account, it would be seen that the wormholes connecting them may crisscross the universe quite densely.

This thought has stimulated the imaginative faculties of astronomers such as Carl Sagan (1934-). It is impossible to think of any way of keeping any sizable piece of matter intact as it approaches a black hole, let alone having it pass intact through a wormhole and out the white hole, yet Sagan does not allow that to limit his speculations.

After all, we can do things that to our primitive forebears would seem inconceivable, and Sagan wonders if an advanced civilization might not devise ways of blocking off gravitational and tidal effects so that a ship may make use of wormholes to travel vast distances in a moment of time.

Suppose there were an advanced civilization in the universe right now that had developed a thorough map in which the wormholes were plotted with their black-hole entrances and their white-hole exits. The smaller wormholes would be more numerous, of course, and therefore more useful.

Imagine a cosmic empire threaded together through a network of such wormholes, with civilized centers located near the entrances and exits. It would be as important, after all, for a world to be located near a transportational crossing point of this sort as it is for an Earth city to be built at some harbor or some river ford.

The planets nearest the tunnels might be a safe distance away, but nearer still would be enormous space stations built as bases for the ships moving through the tunnels and as power stations for the home planets.

And how does the wormhole theory affect the past and future of the universe?

Even though the universe is expanding, is it possible that the expansion is balanced by matter being shifted into the past through the wormholes?

Certainly the dozens of quasars we have detected are all billions of light-years away from us, and we see them, therefore, as they were billions of years ago. Furthermore, they are heavily weighted toward the greater distances and more remote past. It is estimated that if quasars were evenly spaced throughout the universe, there would be several hundred of them nearer and brighter than 3C273, which is the nearest and brightest now.

Well, then, do we have an eternal universe, after all, a kind of continuous creation in another sense?

Has the universe been expanding for countless eons, through all eternity in fact, without ever having expanded beyond the present level because the wormholes create a closed circuit, sending matter back into the more contracted past to begin expansion all over?

Has the universe never really been entirely contracted, and has there never really been a big bang? Do we think there was a big bang only because we are more aware of the expansion half of the cycle involving the galaxies and are not aware of matter sweeping back through wormholes?

But if there was no big bang, how do we account for the background radiation that is the echo of the big bang? Can this radiation be the product of the overall backward flow of matter into the far past? Can the white holes or quasars be numerous "little bangs" that add up to the big bang and produce the background radiation?

And if all this is so, where does the energy come from that keeps the universe endlessly recycling? If the universe runs down as it expands (this is referred to as an *increase of entropy* by physicists), does it wind up again (*decreasing entropy*) as it moves back in time through the wormholes?

There are no answers to any of these questions at present. All is speculation, including the very existence of wormholes and white holes.

THE COSMIC EGG

It must be admitted that the notion that the universe is continually recycling is a rather tenuous speculation.

If we dismiss it, however, we are left with the big bang—either as a one-time affair if we are living in an open universe, or as an endlessly repeated phenomenon if the universe is closed and oscillating. Either way there is a problem. What is the nature of the cosmic egg?

When the cosmic egg was first suggested, it was viewed very much as we now view neutron stars. The trouble is that a cosmic egg with all the mass of the universe (equal to the mass of 100,000,000,000 galaxies, perhaps) is certainly too large to be a neutron star. If it is true that anything with more than 3.2 times the mass of our Sun must form a black hole when it collapses, then the cosmic egg was the biggest of all black holes.

How, then, could it have exploded and yielded the big bang? Black holes do not explode.

Suppose we imagine a contracting universe, which would form black holes of varying sizes as it contracts. The individual black holes might bleed away some of their mass through wormholes, counteracting the overall contraction but not by enough to stop it altogether (or neither the expanding universe nor we would be here today).

As the universe compresses, the black holes grow at the expense of non-black-hole matter and, more and more frequently, collide and coalesce. Eventually, of course, all the black holes coalesce into the cosmic egg. It loses matter through its wormhole at an enormous rate, producing the biggest conceivable white hole at the other end. It is the white hole of the cosmic egg, then, that was the big bang that created our expanding universe. This would hold good whether the universe is open or closed, whether the cosmic egg is formed only once or repeatedly.

Of course, this solution will only work if wormholes and white holes truly exist, which is uncertain. And even if they do exist, it will only work if the cosmic egg is rotating. But is it?

There is certainly angular momentum in the universe, but it could have been created, despite the conservation law, where none had earlier existed.

That is because there are two kinds of angular momentum, in opposite sense. An object can rotate either clockwise or counterclockwise (positive or negative if you prefer). Two objects with equal angular momentum, one positive and one negative, will, if they collide and coalesce, end with zero angular momentum, the energy of the two rotary motions being converted into heat. In reverse, an object with zero angular momentum can, with the addition of appropriate energy, split to form two subobjects, one with positive angular momentum and the other with negative angular momentum.

The objects in the universe may all have angular momentum, but it is very likely that some of that angular momentum is positive and some negative. We have no way of knowing whether one kind is present in greater quantities than the other. If such lopsidedness does exist, then when all the matter of the universe collapses into a cosmic egg, that cosmic egg will end up with an amount of angular momentum equal to the excess of one kind over the other.

It may, however, be that the amount of angular momentum of one kind in the universe is equal to the amount of the other kind. In that case, the cosmic egg, when it forms, will have no angular momentum, and will be dead. We can't rely on wormholes and white holes for the big bang, then.

What else?

Just as angular momentum of two opposite kinds exist, so matter of two opposite kinds exists.

An electron is balanced by an antielectron, or positron. When an electron and a positron combine, there is a mutual annihilation of the two particles. No mass at all is left. It is converted into energy in the form of gamma rays. In the same way, a proton and an antiproton will combine to lose mass and form energy; and so will a neutron and an antineutron.

We can have matter built up of protons, neutrons, and electrons; and antimatter built up of antiprotons, antineutrons, and antielectrons. In that case any mass of matter combining with an equal mass of antimatter, will undergo mutual annihilation to form gamma rays.

In reverse, mass can be formed from energy, but never as one kind of particle only. For every electron that is formed an antielectron must be formed, for every proton an antiproton, for every neutron an antineutron. In short, when energy is turned into matter, an equal quantity of antimatter must also be formed.

But if that is so, where is the antimatter that must have been formed at the same time that the matter of the universe was formed?

The Earth is certainly entirely matter (except for vanishingly small traces of antimatter formed in the laboratory or found among cosmic rays). In fact the whole solar system is entirely matter, and in all probability so is the entire galactic unit of which we are part.

Where is the antimatter? Perhaps there are also galactic units that are entirely antimatter. There may be galactic units and antigalactic units, which because of the general expansion of the universe never come in contact and never engage in mutual annihilation. Just as matter forms black holes, antimatter will form anti-black holes. These two kinds of black holes are in all respects identical except for being made up of opposite substances.

If the universe was ever, in the past, contracting, black holes and anti-black holes formed even more easily; and as contraction continued, the chances of collision between two black holes of opposite nature and a consequent enormous mutual annihilation increased. In the final coalescence there was the greatest of all great mutual annihilations.

The total mass of the universe disappeared and with it the gravitational field that keeps the black hole, and the cosmic egg for that matter, in existence. In its place was incredibly energetic radiation, which expanded outward. That would be the big bang.

Some period after the big bang the energy, becoming less intense through expansion, would be tame enough to form matter and antimatter once more—the two forming separate galactic units by some mechanism that, it must be admitted, has not been worked out—and the expanding universe would take shape.

From this view of the big bang as the mutual annihilation of matter and antimatter, it doesn't matter whether the cosmic egg is rotating or not, or whether it is alive or dead.

Yet we have no evidence that there exist antigalactic units. Can it be that for some reason we do not as yet understand the universe consists simply of matter?

We might argue that this is impossible; the universe cannot consist simply of matter, as that would make the big bang impossible. Or we might think of a way of accounting for the big bang even in a universe of matter only, and even if, on contracting, that universe forms a cosmic egg that is not rotating and is therefore a dead black hole.

Well, according to the equations used to explain the formation of black holes the size of the Schwarzschild radius is proportional to the mass of the black hole.

A black hole the mass of the Sun has a Schwarzschild radius of 3 kilometers and is therefore 6 kilometers across. A black hole that is twice the mass of the Sun is twice as large across —12 kilometers. However, a sphere that is twice as large across as a smaller sphere has eight times as much volume as the smaller sphere. It follows that a black hole with twice the mass of the Sun has that twice the mass spread over eight times the volume. The density of the larger black hole is only one-fourth the density of the smaller black hole.

In other words, the more massive a black hole is, the larger and the less dense it is.

Suppose our entire galaxy, which is about 100,000,000,000 times the mass of our Sun, were squeezed into a black hole. Its diameter would be 600,000,000,000 kilometers, and its average density would be about 0.000001 grams per cubic centimeter. The galactic black hole would be more than 50 times as wide as Pluto's orbit and would be no more dense than a gas.

Suppose that all the galaxies of the universe, possibly 100,000,000,000 of them, collapsed into a black hole. Such a black hole, containing all the matter of the universe, would be 10,000,-000,000 light-years across, and its average density would be that of an exceedingly thin gas.

Yet no matter how thin this gas, the structure is a black hole.

Suppose the total mass of the universe is 2.5 times as large as it seems to astronomers to be. In that case the black hole formed by all the matter of the universe is 25,000,000,000 light-years across, and that happens to be about the diameter of the actual universe we live in (as far as we know).

It is quite possible, then, that the entire universe is itself a black hole (as has been suggested by the physicist Kip Thorne).

If it is, then very likely it has always been a black hole and will always be a black hole. If that is so, we live within a black hole, and if we want to know what conditions are like in a black hole (provided it is extremely massive), we have but to look around.

As the universe collapses, then, we might imagine the formation of any number of relatively small black holes (black holes within a black hole!) with very limited diameters. In the last few seconds of final catastrophic collapse, however, when all the black holes coalesce into one cosmic black hole, the Schwarzschild radius springs outward and outward to the extremity of the known universe.

And it may be that *within* the Schwarzschild radius there is the possibility of explosion. It may be that as the Schwarzschild radius recedes billions of light-years in a flash, the cosmic egg at the very instant of formation springs outward to follow, and *that* is the big bang.

If that is so, we might argue that the universe cannot be open whatever the present state of the evidence, since the universe cannot expand beyond its Schwarzschild radius. Somehow the expansion will have to cease at that point, and then it must inevitably begin to contract again and start the cycle over. (Some argue that with each big bang, a totally different expanding universe with different laws of nature gets under way.)

Can it be, then, that what we see all about us is the unimaginably slow breathing cycle (tens of billions of years in and tens of billions of years out) of a universe-sized black hole?

And can it be that separated from our universe in some fashion we cannot as yet grasp, there are many other black holes of various sizes, perhaps an infinite number of them, all expanding and contracting, each at its own rate?

And we are in one of them—and through the wonders of thought and reason it may be that from our station on a less-than-dust speck lost deep within one of these universes we have drawn ourselves a picture of the existence and behavior of them all.

APPENDIX 1

EXPONENTIAL NUMBERS

NUMBERS CAN, FOR convenience, be written as multiples of 10. Thus, $100 = 10 \times 10$; $1,000 = 10 \times 10 \times 10$; $1,000,000 = 10 \times 10 \times 10 \times 10 \times 10 \times 10$; and so on. A short way of writing such numbers is to indicate the number of 10s involved in the multiplication as a small number (or "exponent") to the upper right of the 10.

Thus, if $100 = 10 \times 10$, we can say that $100 = 10^2$. In the same way $1,000 = 10^3$ and $1,000,000 = 10^6$. As it turns out, the exponent equals the number of zeroes in the large number. The number $1,000,000,000,000,000,000,000,000,000,000,000,000$ (a trillion trillion trillion) has 36 zeroes in it and can be written 10^{36}.

The exponential system also works for fractions. The number $1/100$ is $1/10^2$, and there are sound algebraic reasons for writing it as 10^{-2}. In the same way $1/1,000 = 1/10^3 = 10^{-3}$ and

235

$1/1,000,000 = 1/10^6 = 10^{-6}$. If you write such a number in decimals, the exponent is always one greater than the number of zeroes. Thus $1/1,000,000 = 0.000001$, where five zeroes are to the right of the decimal point, so that the exponential figure is 10^{-6}. If you want to count the single zero usually put to the left of the decimal point, the exponent equals the number of zeros.

Thus 0.00000000000000000000000000000000-00001 (or one-trillionth-trillionth-trillionth) is 10^{-36}.

If you have a number such as 6,000,000, it is equal to $6 \times 1,000,000$ or 6×10^6. In the same way 45,200,000 is equal to $4.52 \times 10,000,000 = 4.52 \times 10^7$. And 0.000013 is equal to $1.3 \times 0.00001 = 1.3 \times 10^{-4}$.

APPENDIX 2

THE METRIC SYSTEM

THE METRIC SYSTEM builds up measurements in units of ten exclusively, as opposed to the common system where the buildup can be by almost any digit, so that twelve inches make a foot, three feet make a yard, and 1,760 yards make a mile.

The metric unit of distance is the meter, and it can be built up (or down) by stages of ten through a series of prefixes:

1 kilometer = 10 hectometers = 1,000 meters
1 hectometer = 10 dekameters = 100 meters
1 dekameter = 10 meters
1 meter
1 decimeter = 1/10 meter
1 centimeter = 1/10 decimeter = 1/100 meter
1 millimeter = 1/10 centimeter = 1/1,000 meter

There are other prefixes, too, both larger and

smaller. For instance, a megameter is 1,000 kilometers; and a micrometer is 1/1,000 millimeter, but these are rarely used.

This simple use of 10 as stages of measurement makes the metric system far more simple to learn and use than the common system. As a result the entire civilized world, except the United States, has gone metric. Someday we will too.

Since a meter is equal to 39.37 inches, or 1.094 yards, a kilometer is 1,094 × 1,000 or 1,094 yards. This comes to 0.621 miles or just about five eighths of a mile. A centimeter, on the other hand, is 39.37 divided by 100 or 0.3937 inch—or just about two fifths of an inch.

A cubic centimeter is 2/5 × 2/5 × 2/5 or 0.064 cubic inch—which makes it about one sixteenth of a cubic inch.

Another metric measure used in the book involves mass. Here the basic unit is the gram. This is a rather small unit, since it is equivalent to 0.035 ounce, or a little more than a thirtieth of an ounce. A kilogram = 1,000 grams = 0.035 × 1,000 ounces = 35 ounces or about 2.2 pounds.

Other more complicated measurement units, such as "dynes" (mentioned early in the book), can be built up out of centimeters, grams, and seconds. Time units such as seconds, minutes, days, and years remain the same in the metric system as in the common system.

APPENDIX 3

TEMPERATURE SCALES

THE ENTIRE CIVILIZED world, except the United States, uses the Celsius temperature scale in which the freezing point of water is set at 0 degrees and the boiling point of water at 100 degrees. The United States will some day adopt this system but as of now it still uses the Fahrenheit scale in which the freezing point of water is set at 32 degrees and the boiling point of water at 212 degrees.

The number of degrees between freezing and boiling of water is 100 degrees in the Celsius scale and 180 degrees in the Fahrenheit scale. Therefore a Celsius degree is 180/100 or 9/5 the size of a Fahrenheit degree. Conversely a Fahrenheit degree is 100/180 or 5/9 the size of a Celsius degree.

You cannot convert one into the other just by taking into account the difference in size of degrees. There is also the question of the different position of the zero, which in the Celsius scale is at the freezing point of water, and in the Fahrenheit scale is 32 Fahrenheit degrees below the freezing point of water.

To convert, then, you must use the following equations:

degrees F = 9/5 (degrees C) + 32
degrees C = 5/9 (degrees F) − 32

INDEX